NOISE AND OTHER
INTERFERING SIGNALS

NOISE AND OTHER INTERFERING SIGNALS

Ralph Morrison

A WILEY-INTERSCIENCE PUBLICATION

JOHN WILEY & SONS, INC.

New York / Chichester / Brisbane / Toronto / Singapore

Copyright © 1992 by John Wiley & Sons, Inc.

Library of Congress Cataloging in Publication Data:
Morrison, Ralph.
   Noise and other interfering signals / Ralph Morrison.

      p.   cm.
   "A Wiley-Interscience publication."
   Includes bibliographical references and index.
   1. Electronic noise.   2. Shielding (Electricity)
3. Electromagnetic compatibility.   I. Title.

TK7867.5.M64   1991   621.382′24—dc20

ISBN 0-471-54288-1   91-15226
                                         CIP

Printed in the United States of America

10 9 8 7 6 5 4 3 2

PREFACE

Noise and interference affect all electrical engineering. In mechanical engineering the problems of fit, roughness, temperature coefficient, and accumulative error play an important role. Similarly in electrical engineering limitations of noise, linearity, accuracy, drift, crosstalk, and radiation all affect performance. The engineer spends much of his time worrying about these parameters. The design itself is often very easy. Making it work in the field is hard.

Some noise factors are intrinsic to the devices selected to do the job. In most cases the worst enemy is the noise brought in by the environment that surrounds the equipment. The environment includes association with power lines, earth, equipment, cables, and external transmitters. The constant battle is how to make equipment work well in each and every situation.

The battle is not won by using words. It is best approached by applying the basic understanding afforded by physics. Circuit theory has limited application because most noise phenomena can best be explained using a bit of field theory. The explanations never have to resort to solving difficult differential equations, but only to basic ideas. Engineers often find it hard to leave their well-learned circuit theory, even though it hardly fits the occasion.

Words do play a complex role in understanding. The power-engineering definitions are alien to most electronics people. The word ground is very general to a circuit engineer, but very specific to a power engineer. As a consequence they do not communicate with each other. Engineers invent words like clean ground or digital ground, concepts not taught in any textbook. They are an invention made in the frustration of trying to find an answer to the noise problem.

Circuit engineers can fight their way through to a solution on a circuit board, but when a building or a room full of equipment is involved, the oscilloscope and voltmeter are inappropriate. Try to measure the frequency response of a ground plane or the impedance of a steel girder and the problem becomes very clear.

In a specific sense, noise is the random motion of electrons. In the broadest sense, it is any error that affects the answer in a measurement or electrical process. This of course leads to the ground plane, the utility power, the nearby radio station, the flaws or limitations in electronic design, and the misconceptions that are ever present in the engineering community. In other words, the subject of noise touches upon many topics, including human behavior.

No one person has the experience to write about every phase of noise. This book represents the experience gained by one person from designing and building instrumentation and then making it work in the field. It comes from experience gained consulting in many different facilities. The author has taught grounding and shielding as a topic to thousands of students and has consulted in dozens of facilities. The sum of the experience is represented in this book. The acceptance by engineering students of the material taught, in a sense, adds credence to the material and to the teaching approach.

The reader must let go of a few of his or her pet ideas to use this book. These pet ideas are hard to locate and only the reader can find them through introspection. If you need this book, yielding for a moment is the best way to learn in this rather mixed field.

I want to thank the many readers of my previous books. It is a bit presumptuous to write a book on such an elementary subject as noise. The success of previous efforts was a spur to taking pen (computer) in hand again. Time will tell just how well this effort is received.

I also wish to thank George Telecki of John Wiley & Sons for the vote of confidence that is represented by suggesting that this book be written.

RALPH MORRISON

Pasadena, California
August 1991

CONTENTS

2 Power Supplies   23

3 Transmission Lines in Analog and Digital Transmission   45

8   Large Industrial Systems   **123**

NOISE AND OTHER
INTERFERING SIGNALS

1

NOISE

1.1 INTRODUCTION

Noise takes on many forms. Noise can be acoustic, electrical, or even optical. The word has a negative connotation that implies something undesirable. Noise to some implies something random, but to others it is simply an undesirable phenomenon. Coughing during a concert is noise because it is not organized to fit the music. Traffic is noise that interferes with a normal conversation. An interfering radio signal is noise to the correct signal.

There are certain noise signatures that are expected and representative of good operation. When an instrument performs at its theoretical noise limit, this is certainly not a negative attribute. When good information can be pulled from a noisy signal, then the noise is not a problem.

In electronic equipment there are specific sources of internal interference that can be eliminated by a proper design. These noises can be electrical in origin, such as a power-supply coupling, a noisy component, or transistor noise. These noises can also be seen, heard, or sensed by humans or other pieces of electronics. Other noises are coupled into the equipment from outside sources, such as the arcing of switches, radio signals, lightning pulses, or power-line disturbances. The range of noise problems is significant and is the subject of this book.

1.2   ACCEPTABLE LIMITS OF NOISE

Signals and their associated noise levels can vary greatly and still be acceptable. The eye is very discerning and a single spot on a lace curtain is disturbing. If the idea is to recognize that it is a lace curtain, the spot is of no significance. A good measure of noise sensitivity can be illustrated by considering a video signal. If the signal is larger than the random noise by a factor of 100, the picture will show some graininess or snow. If the noise is one tenth of the signal, the electronics may not be able to provide a stable picture.

The ear is often far less discerning than the eye. Random noise that is one tenth of the music level may not be heard until the music is turned off. If the noise is in the form of flutter or wow where the pitch is slowly changed, the mind can sense a very small amount of this phenomenon and it can be very upsetting. The disturbance results because the basic information is modified. A constant buzz or hum buried in the music can be very disturbing to some listeners or simply cancelled by the mind for others. When there are no other sounds present, the continuous buzzing of a fly can be very maddening.

Digital signals by their very nature are large signals. Information is contained in literally millions of voltage states. Information is moved by making millions of voltage transitions per second. The structure of the information is such that the error levels must be essentially zero. One error in 100 million operations is simply not acceptable, because commands, numbers, letters, and addresses all look the same. An error that changes a command or an address stops the entire digital process. This means that spikes or glitches that alter the data at an inopportune moment are unacceptable.

The human mind can fill in blank spots, but a computer can not. An ink spot on a printed page may still allow the page to be read. Even leaving out all the "e's" may still leave the information in readable form. Certainly a misspelled word is not a problem. A misspelled computer direction is not acceptable.

Noise can take on many forms. It can be continuous in nature, such as a power hum. It can be random like the fluttering of leaves on a tree or it can be transient like a bolt of lightning or the operation of a power switch. The impact that any noise has is dependent on the device involved. Humans have certain sensitivities, as do computers or meters. Engineers have categorized noise in many ways. White noise is a particular form of random noise. Spurious radio signals are a form of coherent noise. In order to deal with noise these terms must be clearly understood. To understand the impact of noise, the sensitivity levels must be understood.

1.3   THE MEASURE OF VARIOUS TYPES OF SIGNALS

Noise voltage or current measurements must be understood before further discussion is useful. A current flowing in a resistor converts this current to a voltage. Once the measure of voltage is understood, a current measurement follows immediately by Ohm's law.

Voltage is a measure of work required to move a small unit of charge between two points in an electromagnetic field. Most of us would rather place a voltmeter between these two points and accept the meter reading rather than trouble over this definition. There are times when this simple procedure gives misleading information, such as when the meter leads couple to a changing magnetic field. At frequencies below 100 kHz the readings in most applications are apt to be fairly reliable without regard to field coupling.

The simplest potential to measure is a steady electric pressure, such as from a battery, using a voltmeter. The battery forces current in a small coil. This current forms a magnetic field that rotates the coil against a spring. A needle mounted on the coil indicates the size of this motion and reads the voltage. This static voltage is said to be dc, which stands for direct current. This might be a misleading abbreviation because the measure is of a static voltage not of a current.

The usual measure of voltage is its heating effect in a known resistor. If one voltage heats the same as another, the two voltages are equal in a heating sense. This idea of heating is important when there is a need to measure a nonstatic voltage. If a changing but repetitive voltage heats a resistor the same as a corresponding dc value, the two voltages are given the same measure. It is very convenient to give a sine wave a measure of 10 V if the voltage heats a resistor the same as a 10-V battery.

A static voltage dissipates power in a resistor given by

$$V^2/R \tag{1}$$

When a sinusoidal voltage is involved that has a peak of V volts the power dissipation is given by

$$V^2/2R \tag{2}$$

The power is less because part of the time the voltage is low and does little heating. In a heating sense this sinusoid is a smaller voltage. If the dc value is adjusted downward until it heats the same as a 10-V peak sine wave, the voltage is 0.707 times the peak value. In other words, a 10-V peak sine wave has a heating voltage of 7.07 V. More accurately, the voltage is 10 V divided by the square root of 2.

For a general waveform that is repetitive in character over a time period T, the equivalent heating voltage V is given by the equation

$$V = \left[1/T \int_0^T A^2(t)\, dt \right]^{1/2} \tag{3}$$

where $A(t)$ is its amplitude as a function of time. This heating measure of voltage is called the rms or root–mean–square value. This description comes directly from the square root of the squared term in Equation (3).

Figure 1.1 indicates the rms value for a group of repetitive voltage waveforms. The waveforms are drawn so that the rms amplitudes are constant.

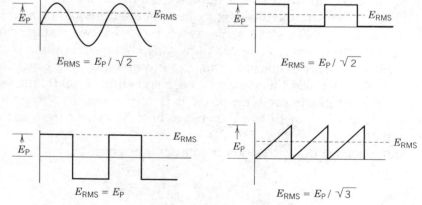

FIGURE 1.1   The rms value for a group of waveforms.

1.4 NOISE MEASUREMENTS

Noise by its very nature might come in bursts or as a random event. If the noise is viewed as a changing voltage, the only accurate measure that makes sense is a detailed description of the voltage waveform. Sometimes an envelope of the voltage waveform provides a good measure. At other times a measure of the peak voltage is sufficient. Noise that is not predictable cannot be measured by using a simple voltmeter.

Many sources of noise are random yet very repetitive in character over a period of time. Some examples might be the noise of a steady rainfall on a roof, the random way electrons reach a collector in a transistor, or the way gas particles hit a wall. These signals can be given an rms measure because the signal for any one sample period is the same as any other sample period within a small error band. Electrical noise that has this quality is said to have an rms value.

1.5 NOISE QUALITY OR CHARACTER

Random noise processes vary in quality. The sound of a crowd, the sound of rain drops, the roar of a jet engine, or the sound of rustling leaves all have a different signature. Something in the air-pressure waveform that reaches our ear is different. If these air-pressure waveforms are converted to a voltage by a microphone, the voltages will also have a different signature. The voltage from the microphone will be very repetitive, yet each signal will have a different measure. This measure is its frequency spectrum. To understand the ideas of a frequency spectrum the nature of various waveforms needs to be analyzed.

1.6 WAVEFORMS AND THEIR FREQUENCY CONTENT

The simplest signal turns out to be a sinusoid. Sinusoidal motion is found in nature all the time. The motions of a mass and spring, the phases of the moon, the length of the day over a year, the motion of a tuning fork, and the swing of a pendulum are a few sinusoidal motions.

It is interesting to see what happens when three sinusoids are added together. If the periods are simple multiples of each other, the resulting waveform repeats over and over again. This is illustrated in

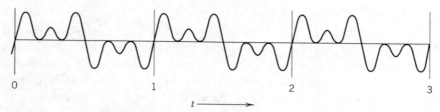

FIGURE 1.2   The sum of three sine waves.

Figure 1.2, where the sine waves are sin ωt, sin $3\omega t$, and sin $5\omega t$ and $\omega = 2\pi f$ and $f = 1$.

If t represents time, then when $t = 0$ all three sine waves are zero. As t increases from 0 to 1 the first sinusoid goes through one cycle. During this time sin $3\omega t$ goes through 3 cycles and sin $5\omega t$ goes through 5 cycles. The sum of these three signals repeats exactly when t increases from 2 to 3, from 3 to 4, or between any two adjacent integers.

It turns out that any repetitive waveform can be constructed out of sine waves. Consider the sum of the sine waves sin ωt, $\frac{1}{3}$ sin $3\omega t$, $\frac{1}{5}$ sin $5\omega t$, and $\frac{1}{7}$ sin $7\omega t$, where $\omega = 2\pi f$. The resulting waveform is very nearly a square wave, as shown in Figure 1.3. This square wave repeats once per second. If all of the sine waves implied in the sequence are supplied, the waveform is exactly square.

Since a series of sinusoids makes up a square wave, it is only correct to say that a square wave is made up of a series of sinusoids. These sinusoids are said to be harmonics of the fundamental sine wave. For example $\frac{1}{3}$ sin $3\omega t$ is said to be the third harmonic of sin ωt. The square wave is said to have a spectrum that consists of odd harmonics of the fundamental sinusoid. The fundamental sinusoid in this example repeats every second when $\omega = 2\pi$. This is the same as saying the sinusoid has a frequency of 1 Hz. The third harmonic thus has a period of

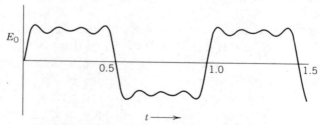

FIGURE 1.3   The formation of a square wave.

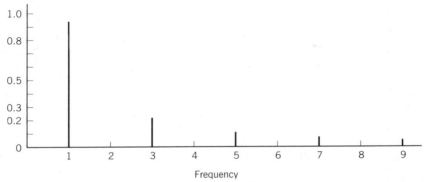

FIGURE 1.4   The spectrum of a square wave.

$\frac{1}{3}$ second or a frequency of 3 Hz. The frequency spectrum of a square wave is shown in Figure 1.4. The square wave, if used to produce sound, would have the same pitch as the fundamental sine wave, but it would have a different character or timbre. The spectrum thus tells us what this character is.

1.7   THE SPECTRUM OF A SINGLE EVENT

The spectrum of a repeated set of pulses, where the period is long and the individual pulses are short, is shown in Figure 1.5. The long period requires a low fundamental frequency and the short pulses require many closely spaced harmonics.

It is interesting to take this repeated set of pulses to the limit by increasing the time between pulses to infinity. When this happens the harmonics become continuous so that there is signal content at every frequency. The amplitude at each frequency must approach zero, as an infinite set of sinusoids would involve an infinite signal.

The only way this spectrum can be handled is to describe the voltage over a portion of the bandwidth. The individual harmonics have lost their meaning. Figure 1.6 shows the spectrum of a single event where the amplitude is indicated as voltage per unit bandwidth. In this case the unit of bandwidth is 1 Hz. The voltage used to describe the signal in this band would heat the same as a corresponding dc voltage.

The unit of bandwidth is arbitrary and represents the amplitude density at a point in the spectrum. The units can be volts per hertz or

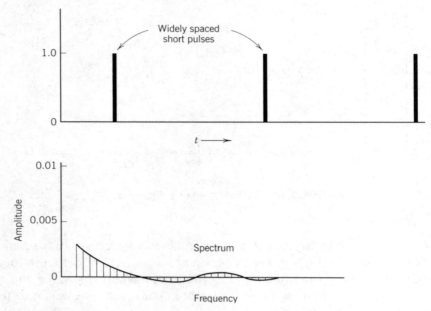

FIGURE 1.5 The spectrum of a repeated set of short pulses.

volts per megahertz. The unit of bandwidth does not necessarily infer that the noise is constant over this bandwidth. For example, if the noise is 10 mV/MHz at 30kHz it might be 20 μV/MHz at 100 kHz.

1.8 PULSE SPECTRUM

Noise that has the same voltage per unit of bandwidth in every part of the spectrum is said to be white noise. This term arises from the idea of white light, which contains all of the colors in equal amounts. The idea is quite general and requires that there be noise at all frequencies as well as dc. In practice, the noise need only have bandwidth that exceeds the range of interest to be considered white noise.

A very wide spectrum can be generated by a single pulse. If the pulse is ideal and lasts zero time with infinite amplitude, and the product of volts times time is unity, the spectrum is that of white noise. This ideal pulse cannot be built any more than a device with infinite bandwidth can be constructed. If a pulse is to be practical and excite a

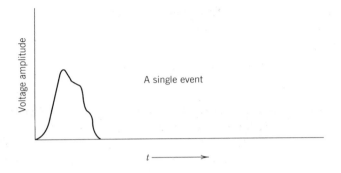

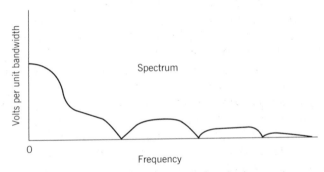

FIGURE 1.6 The spectrum of a single event.

circuit over its entire spectrum, it must have a rise time that is shorter than the rise time of the circuit under test. In this way the pulse is ideal enough.

1.9 THE VOLTAGE OF COMBINED SIGNALS

When sine waves of the same frequency and phase are combined in the same circuit, they add together. A 40-V sinusoid at 10 Hz adds to a 30-V sinusoid at 10 Hz to make a total of 70 V. If the frequencies are not the same, the combined voltage is no longer additive. In general, when the frequencies differ, the total heating equivalent is found by taking the square root of the sum of the voltages squared. In the example given above, if the frequencies differ, the voltages add together to give a total of 50 V. The sum of voltages is given by Equation (4):

$$V = \sqrt{V_1^2 + V_2^2 + V_3^2 \cdots} \qquad (4)$$

When the signal involved is noise, the same problem exists. Assume white noise so that the voltage in each kilohertz of bandwidth is 10 mV. The noise in 2 kHz bandwidth is not 20 mV. The total noise in this wider bandwidth must be calculated using Equation (4) and is 14.14 mV.

A convenient way to describe noise so that it is easier to scale voltages over any bandwidth is to use units of volts per square-root hertz. To find the noise in any bandwidth, the scaling factor is simply the square root of bandwidth. For example, if the noise of an amplifier is $4\,nV/\sqrt{H_z}$, the noise in a 1-MHz bandwidth is $\sqrt{1{,}000{,}000} \times 4\,nV/\sqrt{H_z}$ or 4 μV. This procedure automatically takes into account the fact that noise voltages add per Equation (4). This assumes of course that the noise voltage is constant over this bandwidth.

If the units of noise are given in volts per megahertz, then the noise in a new bandwidth can be calculated by multiplying by the square root of the bandwidth ratio. For example, the noise in a 10-MHz bandwidth is 3.16 times the noise in a 1-MHz bandwidth. This assumes again that the noise figure is constant over this range. If this constancy is not present, then Equation (4) must be applied.

1.10   COHERENT NOISE

Many types of interference are classed as coherent. A signal that can be predicted is said to be coherent. This includes a radio-frequency (rf) carrier, hum from a power supply, or a digital clock. A single pulse is said to be coherent in that the spectrum that makes up the pulse is fully predictable. Pulses arise from lightning, power switching, Electrostatic Discharge (ESD), and power surges. Random noise is said to be incoherent in that the next value is not predictable. Random noise includes noise from transistors and resistors.

When coherent noise couples to a circuit via several paths, the results must be considered additive. This worst-case approach is the safe way to view this type of coupling.

1.11   FORM FACTOR

Form factor is a term used to describe the ratio between two measures of a signal. A sinusoid has a form factor of 1.414, because this is the ratio of rms value to peak value. A series of pulses might have a peak to rms ratio of 10:1. Random white noise has a peak value roughly three times its rms value. Peak values of white noise can theoretically be very large, but for this type of noise 95% of the peaks will fall within this factor of 3.

1.12   VOLTMETERS

The simplest voltmeters average the signal they measure and indicate an rms value. This type of meter is moderately accurate for form factors less than 3. More accurate digital rms meters are available that sample the waveform and calculate the correct rms value. These meters are obviously limited in their sampling rate and cannot handle signals above a certain frequency. Meters that sample data must have a signal filter that limits the input signal to one half the sampling frequency. If the sampled signal is above this 50% frequency, the output signal will be the difference between the sampled frequency and the sampling frequency. This is known as an aliasing error. If the sampled frequency equals the sampling frequency the output will be dc (see Section 5.25).

Meters that detect high-frequency signals in their probe tip are often peak detection devices. The scale readings can be rms however. It is unusual to find rf signals where the form factor makes these reading inaccurate.

1.13   SOURCES OF NOISE OR INTERFERENCE

Magnetic devices are frequently a source of interference. The most complex and most frequent offender is the local power transformer. The connections to the utility power can introduce interference in a variety of ways. For example, the primary leads can serve as transmission lines for high-frequency energy.

The grounds associated with safety and with power are not at the same potential as the local signal ground. The power lines have both differential and common-mode noise. The magnetic fields associated with the transformer iron can couple to the nearby circuits. Current associated with core saturation can also interfere with circuit operation. All of these points will be covered in the following chapters.

Whenever current in a magnetic device is interrupted, large voltages and arcing can result. These signals can be significant enough to damage circuit elements. Signals can couple both capacitively and inductively. Signals can enter via the power leads, control leads, or even shielded cables. These methods of coupling will also be covered in the following chapters.

Pulse-type noise can arrive from electrostatic discharge or ESD, from lightning or atmospherics, from power-factor correction capacitors being switched, or from load switching. These pulses can be brought to the circuit over any conductor entering or leaving the circuit.

Radiated energy from rf transmitters, television broadcasts, or radar sets can be disruptive. These signals may be out-of-band, but they can be detected or rectified causing in-band signals. These signals can enter over conductors or through apertures in metal enclosures. This type of interference is treated in subsequent chapters.

1.14   SOME DEFINITIONS

There are some overused and abused words in engineering that cannot be avoided. These words mean so many things that they often convey very little. It is important to discuss their usage so that the meaning in context will be clear.

Ground.   To the power industry the word ground implies a conductor that eventually connects to earth or soil. In electronics this is not a requirement, although some grounds are eventually tied to earth. A ground is a reference conductor in a circuit. It can be one side of a power supply, a centertap on a transformer, or the frame of a metal cabinet. There can be many grounds or reference conductors in one circuit or facility. Grounds can even float; that is, they can have little or no association with another circuit.

Isolation.   This word can mean physical or electrical separation, high-impedance connections, or separation from a previous ground. Its meaning will vary depending on context.

Shield. A shield can be magnetic or electrostatic. It can be a braided sheath over a conductor, a piece of metal, a conductor in a transformer, a magnetic can, or even an active circuit.

1.15 GAIN

Electronic circuits are used to amplify and condition electrical signals. Without this conditioning the original signal would not be useful. Conditioning a signal includes filtering, offsetting, and changing impedance or waveform. The most important parameter change is usually voltage gain.

The signals present in an electronic circuit include operating voltages as well as signals. In a dc amplifier there is no way to separate these two parameters. The only way gain can be defined for all cases is to measure the ratio of changing signal at two separate points. If an input signal changes from 1 to 2 V and the output changes from 3 to −3 V, the ratio of change is −6. This means there is a gain of −6. The input could change from −1 to 0 and the gain would still be −6.

When the signal of interest is sinusoidal, this waveform is constantly changing. If the incoming signal is 1 V rms and the output signal is 10 V rms, the gain is 10. If the phase of the signal is reversed, the gain is −10. This assumes the meter does not read any dc offset or bias. Circuits that amplify the changing waveform and ignore the offset or bias are called ac amplifiers. Circuits that amplify both static and dynamic information are called dc amplifiers.

In a typical circuit there is voltage gain from every point in the circuit to every following point in the circuit. If there is overall feedback in the circuit, there is gain between any two points. In a well-designed circuit the gain between most internal points to the output is near zero and the gain from the input to the output is well defined. The internal points in question include power supply voltages, offset or bias circuits, collectors, emitters, and bases.

Interference that enters a circuit where there is no gain cannot disturb the circuit. If the circuit is overloaded, this may not be the case. Interference that enters a circuit at its input will be treated like any other normal signal. There is no way to train a circuit to separate a good from a bad signal. If the interference is out of band, there is a chance of rejecting the interference by implied filtering.

1.16 FEEDBACK

Low-frequency circuits often use feedback to provide a known gain, reduce output impedances, raise input impedances, reduce distortion, increase bandwidth, and reduce phase shift. The presence of feedback greatly influences how noise or interference influences a circuit.

Consider a circuit that has two forward gain elements A_1 and A_2 as in Figure 1.7. The input signal is summed with a fraction β of the output signal. The overall gain of this circuit is

$$G = A_1 A_2/(1 + A_1 A_2 \beta) \tag{5}$$

where β is usually formed by a resistor attenuator. If the product $A_1 A_2$ is large compared to unity, the gain G will be very close to

$$G = \left(\frac{1}{\beta}\right) \tag{6}$$

For example, if $\beta = 1/10$ and $A_1 A_2 = 10,000$, the gain G will be near 9.99. If $A_1 A_2$ is 100,000, the gain G will be 9.999. The larger $A_1 A_2$ becomes, the closer the gain will be to its ideal value of $1/\beta$ or 10.000.

1.17 INTERFERENCE AND GAIN

Assume an error signal E enters the circuit in Figure 1.7. The gain to this signal is

$$G = A_2/(1 + A_1 A_2 \beta) \tag{7}$$

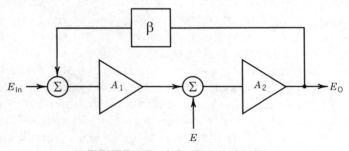

FIGURE 1.7   A feedback circuit.

If $A_1 A_2$ is large, this gain approaches

$$G = 1/\beta \cdot 1/A_1 \tag{8}$$

This error signal will be multiplied by the gain of the circuit and reduced by the gain preceding the point of entry. If A_1 is unity, there is no reduction.

These arguments illustrate how feedback can be used to reduce interference. If there is sufficient gain, the signal is multiplied by $1/\beta$, which is not dependent on the nature of the electronics providing that gain. The only exception is the input circuit where there is no preceding gain. Errors or interference that couple in at the input are multiplied by the gain $1/\beta$.

1.18   THE PENALTIES OF FEEDBACK

Feedback allows the gain to be determined by $1/\beta$ as long as the product $A_1 A_2 \beta$ is greater than unity. This means that the bandwidth of A_1 and A_2 alone can be lacking, yet the feedback circuit will have adequate bandwidth. This can only be true provided that $A_1 A_2$ has a controlled attenuation slope with frequency. If the slope becomes too steep, there will be excess phase shift. At 180° of phase shift the sign of $A_1 A_2$ is reversed. If this occurs at a frequency where $A_1 A_2 \beta$ is greater than unity, the feedback system will be unstable. If the design is marginal, the circuit will have a large overshoot to a step-function input signal. The amplitude response will also show a high gain at the upper band edge.

Circuits with feedback will exhibit a low output impedance. As the frequency increases, the amount of feedback available to lower the output impedance drops. This means that the output impedance rises with frequency. This is the same as an inductance. When this circuit is asked to drive a capacitive load, a series resonant circuit is set up. Under some conditions the circuit can again become unstable. The stability of the capacitively loaded circuit must be analyzed to insure adequate margin. An oscillating circuit can look very much like a noisy circuit under some conditions.

1.19 POWER SUPPLIES

Error signals can enter the signal path from every node in the circuit and this includes the power supplies. The effect is complicated because the power supply is apt to connect to many circuits at the same time. The important entry point is the input circuit as this is where the effect is most felt.

The circuits used at the input are usually differential in character to avoid any sensitivity to power supply change. If the signal gain in the input stage is 100, assume the power voltage gain can be held well below 0.001. Using the ideas just developed a 1-V power supply variation is attenuated by a factor of 100,000. If $1/\beta = 1000$, the output signal would be 10 mV. Referred to the input, this is an error of 10 μV. Stated another way, power supply variations can essentially be eliminated by the use of feedback and the proper design of input stages. Most commercially available ICs or operational amplifiers are designed with well-balanced differential input stages and meet this criteria.

1.20 THE NATURE OF FIELDS

All electrical phenomena are controlled by both electric and magnetic fields. Every voltage has an associated electric field and every current has an associated magnetic field. The electric field starts and ends on charges that reside on the surface of conductors. When a voltage changes in a circuit, the electric field changes. This requires that charges move, which implies a current, which implies a magnetic field. When a current changes, its associated magnetic field changes and this induces voltage into nearby circuits. This voltage implies an electric field.

At frequencies below 100 kHz magnetic processes are usually not dominant. One exception is the magnetic field near a transformer. The electric field provides the basis for understanding noise coupling into analog circuits. Coupling to higher-frequency signals can still be a problem, but this will be treated later.

1.21 THE ELECTRIC OR *E* FIELD

An accumulation of charge on a conductor generates a force field. An example of this might be clothes clinging together after being tumbled

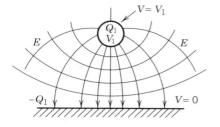

FIGURE 1.8 The electric field between two conductors.

in a dryer. The work required to move a unit of charge from a reference ground to this point of charge accumulation is the voltage at this point. The field can be visualized by drawing lines that follow the direction of the force. By convention every line leaves a unit of positive charge and terminates on a unit of negative charge. The electric field is a force field with an intensity and direction at every point in space. A typical representation is shown in Figure 1.8.

When charge is moved in the electric field, work must be done. This work is stored in the electric field, as it cannot be stored in the conductors. This energy is available to do work such as heat a resistor. This energy is said to be stored in a capacitance. The measure of capacitance is the ratio of charge to voltage. The unit of capacitance is the farad. This is a very large unit and it is more common to use the unit of microfarad (μF) which is one millionth of a farad. A capacitance that stores one microcoulomb of charge per volt has a capacitance of one microfarad.

An electric field or E field exists whenever there are potential differences. When the voltages vary, the field changes and the charges on the conductive surfaces must also change. The field lines in Figure 1.8 concentrate between the two conductors. This defines how the surface current flow is distributed when the voltages change. Current does not use an entire conductor unless the fields terminate on the entire conductor.

1.22 ELECTROSTATIC SHIELDING

The electric field lines in Figure 1.9 are nearly all contained by conductor number 2. A small amount of the field terminates on conductor 3, which is grounded. Grounded in this case means connected to the zero-reference conductor. When the voltage on conductor 1 changes, the charges on all of the conductors must change. The charge

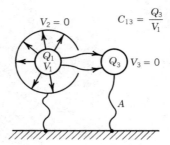

FIGURE 1.9 Mutual capacitance.

moving onto grounded conductor 3 must flow in conductor A. The ratio of charge on conductor 3 to the voltage on conductor 1 is called a mutual capacitance. The notation for a mutual capacitance is C_{13}, while the self-capacitance is noted as C_{11}.

If the field were completely contained by conductor 2, the mutual capacitance would be zero. It would be correct to say that conductor 2 completely shields or guards conductor 1. Since there would be no field external to conductor 1, there would be no work required to move a unit of charge in this region. There would be no charges on the outside of conductor 2.

It is important to realize that all voltages produce electric fields, including signals and power supplies. The fact that some of the voltages are changing simply means that charges are moving. In analog circuits the majority of the electric field energy is confined to the inside of fixed capacitors. The fields that are not contained are controlled by the geometry of the circuit. This geometry controls the nature of coupling to and from external sources. This geometry is the key to understanding interference control.

1.23   THE MAGNETIC OR *H* FIELD

Current flowing in a conductor produces a magnetic field. This field is a force field that has an intensity and a direction at every point in space. One way to detect this force is to place a small magnetic compass in the field. The torque on the needle is a measure of field intensity and the direction the needle points is the direction of the magnetic field. The lines of force form loops that close on themselves. These loops surround every element of current. Another name for these force lines is magnetic flux.

It takes work to establish a current in a conductor. This work is stored as energy in the magnetic field. This energy can be taken out of the field to heat a resistor for example. Just as a capacitance stores electric field energy, an inductance stores magnetic field energy. This energy is often concentrated in a component known as an inductor. The unit of inductance is the henry. By definition a henry stores one unit of magnetic flux per unit of current. The larger the circuit, the more flux is created for the same current. This larger circuit will have a larger inductance.

The H field is constant at a fixed distance from a current carrying conductor. Ampere's law requires that the integral of *H* times distance around a path equals the current *I* that is encircled by the path. For a circular path at a fixed distance from a long conductor carrying a current, Ampere's law requires that

$$2\pi rH = I \tag{9}$$

or

$$H = I/2\pi r \tag{10}$$

This means that *H* is measured in units of amperes per meter.

1.24   THE INDUCTION OR *B* FIELD

When a magnetic field changes, a voltage is induced into any loop of wire that couples to this magnetic flux. The relationship is known as Lenz's law.

$$V = \frac{d\phi}{dt} \tag{11}$$

The flux ϕ that crosses the loop is equal to the induction field intensity times the area of the loop. This induction field is also called the *B* field.

$$B = \phi A \tag{12}$$

The relationship between the induction or B field and the magnetizing or H field is given by

$$B = \mu_0 H \tag{13}$$

where μ_0 the permeability of free space equals $4\pi \cdot 10^{-7}$ and where B is in units of teslas and H is in units of amperes per meter. The tesla is a large unit of measure and is equal to 10^4 gauss. A gauss is the flux density per square centimeter, while a tesla is the flux density per square meter.

For iron that has a relative permeability of μ_r, the relationship between B and H is

$$B = \mu_0 \mu_r H \tag{14}$$

In a typical power transformer B is about 15,000 gauss. If μ_r is 10,000, then by using Equation (14) H must be 119 A/m. If the path length in iron is 0.1 m then the magnetizing current must be 11.9 A. If the primary has 100 turns, the magnetizing current is 0.119 A.

1.25 LENZ'S LAW

When a changing B or induction field couples flux to a conductive loop a voltage results. This induced voltage is given by Equation (15) and makes no mention of permeability. Note that the flux $\phi = BA$ and

$$V = d\phi/dt \tag{15}$$

If the flux threads through n turns, the voltage is simply multiplied by n or

$$V = n d\phi/dt \tag{16}$$

1.26 RADIATION

Electric and magnetic fields both store energy. When a circuit has a static voltage, there is a fixed electric field. When a circuit has a steady current, there is a fixed magnetic field. Most of the time these fields are concentrated in components and only a small fraction of the energy is stored over all space. When the circuit is changing in a dynamic basis the energy stored in components is returned to the circuit. If the energy is dissipated, it is within the circuit itself. Inductive and capacitive reactance imply that energy is stored and returned twice per cycle to every reactive element.

Field energy that is not stored in a component must be stored in space. Field energy cannot be created in zero time nor can it be transported at infinite speed. When the energy is stored locally, there is essentially no time delay in returning the energy to the circuit. When the energy is stored in space, it takes time for the field to propagate out and return to the circuit. The propagation velocity is the speed of light.

At 100 kHz the energy in a field would travel 3000 m in one cycle. In a typical circuit most of the field energy would be located within a distance of 30 m. This means that most of the field energy can return to the circuit in a small fraction of a cycle.

At 10 MHz the energy in a field can travel only 30 m in a full cycle. By the time most of this energy can return to the circuit, the circuit is already generating field energy with the sign of the field reversed. The only field energy that can return to the circuit is located within a few centimeters of the circuit, and this is only a small fraction of the total field energy. In effect the energy that cannot be returned to the circuit is lost or radiated.

Radiation occurs for conductive loops or for straight conductors. A switch closure on a long open conductor might radiate as a dipole antenna. Surge current into a capacitor might radiate as a magnetic loop. The important idea to consider is that this field energy is present in space and is not confined to a component. The nature of a field near a radiator is characterized by the geometry of the radiator. At a distance, the nature of the radiating field is not dependent on the nature of the radiation source except possibly for field polarization.

1.27   THE FAR FIELD AND THE NEAR FIELD

At a far distance from a radiating source the ratio of electric to magnetic field strength is always fixed. In free space the ratio of E to H is always 377 Ω. This ratio is also known as the wave impedance of free space. Given E at 377 V/m, H calculates to be 1 A/m. This number can be assumed for distances greater than a sixth of a wavelength. For example, at 10 MHz this distance is about 5 m. This distance is also known as the near field–far field interface. The near field for magnetic sources is often called the induction field.

The fields that impinge on a circuit as interference can originate from various commercial transmitters or from random sources. The fields that relate to power frequencies or their harmonics are near fields, as the far field–near field interface distance is over 500 miles.

2

POWER SUPPLIES

2.1 POWER SUPPLIES—SINGLE PHASE

The most common type of power supply involves the rectification of utility power. The resulting dc voltages are used to operate various integrated circuits, transistors, or electromechanical devices. The secondary coils of the power transformer provide the voltage for rectification. The usual procedure is to use capacitor input filters as in Figure 2.1.

The transformer supplies current to charge the capacitor when the secondary voltage exceeds the capacitor voltage plus the diode drop. In a typical application the current flows 20% of the time. The load drains charge from the capacitor during the entire cycle, which causes the voltage to sag, except during the charging period. The voltage waveform is shown in Figure 2.2. It is good practice to keep the peak-to-peak ripple voltage below 1 V at full load. This defines the size of

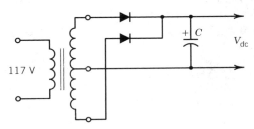

FIGURE 2.1   A full-wave centertap rectifier with a capacitor input filter.

23

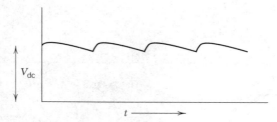

FIGURE 2.2 The voltage waveform of a full-wave centertap rectifier system.

the capacitor. For example a 10-mA load flowing for one half cycle or 8 ms is 80 μC. Since $C = Q/V$ it takes 80 μF to limit the voltage drop to 1 V. There are several potential noise sources in this rectifier system:

1. The peak filter capacitor current can be quite high. This current can cause excessive voltage drop in all conductors associated with the rectifier and capacitor. Load and ground connections should be made at the capacitor, not at the centertap of the transformer, to avoid sensing this potential drop. Heavy traces should be used if the design involves a printed wiring board.

2. The pulses of capacitor current result in magnetic fields. The field is proportional to the loop area carrying this current. This loop area should always be made as small as practical.

3. The two halves of the secondary winding should couple to the entire primary. If this is not done, the leakage inductance of the transformer will be excessive. The peak capacitor current flowing in this inductance will cause excess magnetic fields near the transformer. The voltage regulation will also be poor.

4. The transformer is not efficiently used as each half of the secondary conducts on alternate half cycles. The situation is somewhat improved when a balanced full-wave centertap arrangement is used, as in Figure 2.3. Here both halves of the secondary are used each half-cycle. The problems listed above also apply to this arrangement.

Half-wave rectifiers that allow a dc current to flow in the secondary coils tend to saturate the transformer core. If the power levels are small, the added magnetizing current on the primary may not be troublesome. In a severe case the excess of primary current can distort the voltage waveform and overheat the transformer.

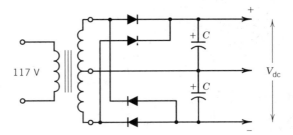

FIGURE 2.3   A balanced full-wave centertap rectifier system.

2.2   THE FULL-WAVE BRIDGE

This rectifier system is shown in Figure 2.4. The absence of a center-tap provides a possible new source of difficulty. The parasitic capaci-tances internal to the transformer are connected to the load through the diodes, not through the centertap as in Figure 2.3. When the diodes conduct the connection is solid. During a large fraction of the cycle the diodes are back-biased and the connection is made through the capacitances of the diodes. In some sensitive circuits this type of in-determinate connection can be troublesome. Full-wave bridge rectifier circuits are less expensive than centertapped rectifiers because a trans-former centertap is not required and only one filter capacitor is used.

2.3   THREE-PHASE RECTIFIER SYSTEMS

Each phase of a three-phase system can be used as a separate source of rectified power. The neutral on the secondary of a transformer

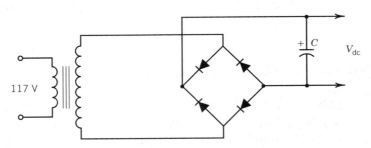

FIGURE 2.4   A full-wave bridge rectifier.

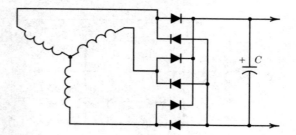

FIGURE 2.5   A full-wave three-phase rectifier system without neutral current flow.

would be the common conductor for each rectifier system. If this transformer is located within commercial equipment, the neutral run can be short and not subject to the grounding (earthing) requirements of the National Electrical Code. The Code requires a new grounding of the secondary neutral for a separately derived system.

Neutral current for a balanced, three-phase resistive load averages to zero. Balanced rectified currents do not flow over the entire cycle and as a result neutral currents do not average to zero. In fact, neutral current can be as high as 1.73 times the line current. When the neutral is used as a reference conductor, the potential drop along its length can appear as interference to some equipment. Because the neutral current has high harmonic content, the inductive effects of a long neutral run are emphasized.

In installations using large amounts of three-phase rectified power the circuits can be designed without requiring current flow in a neutral conductor. This obviously avoids the problems described above. A full-wave three-phase rectifier system is shown in Figure 2.5. The ripple frequency is largely 360 Hz. If six secondary coils are used on the secondary of a three-phase transformer, the ripple frequency becomes 720 Hz. The ripple level can be low enough that filter capacitors may not be required.

2.4   REGULATORS AND STABILITY

Power supply voltages fluctuate with line voltage and load variation. In many applications regulators are needed to remove these supply variations. Regulator circuits are available that hold the voltage or current constant. The basic idea in regulation is to compare the output pa-

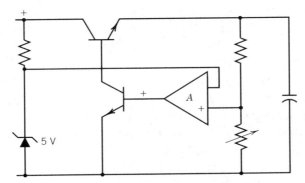

FIGURE 2.6   A feedback voltage regulator.

rameter with a reference value and amplify the error difference. This amplified difference is used to control a series element that regulates the output parameter. These are the basic elements of a feedback control system.

The pass element absorbs the excess voltage when the line voltage rises, and it serves to pass more current when the load resistance is dropped. To be effective, the regulation must be unconditionally stable and quick to respond. When the circuits are improperly built, the problems that result can look just like noise. A typical feedback scheme is shown in Figure 2.6.

A quality voltage regulator has a low output impedance at dc as well as at high frequencies. When the load demands more current in a step manner, the regulator can correct the voltage within microseconds. During this time the added current for the load must come from a local capacitor until the correcting amplifier can respond. If the correcting amplifier overshoots or rings excessively, the regulator is improperly designed.

Many noise problems can arise as a result of an improper regulator design or layout:

1. *The reference voltage is noisy.* When this happens the fluctuations are regulated into the output voltage. No practical amount of bypassing can change the result.

2. *The reference voltage or the input error amplifier couples to some noise or hum.* This signal is regulated into the output voltage and no amount of filtering will remove the problem.

3. *The stability of the regulator is inadequate.* A varying load such as a logic signal or a clock will cause the regulator to overshoot. When this happens this signal is coupled into other circuits and is a type of interference.

4. *The load excursions exceed the regulator capability.* If the load is a step capacitor, the current demand may exceed the regulator's capability. The regulator ceases to function until the capacitor is adequately charged or discharged. Also, some regulators may not work at no load.

5. *The load is dynamic or active.* Current may not be reversible in the pass element. If this demand is made, the regulator will be overloaded. Recovery time may be much longer than the normal regulation response time.

A feedback regulator is fully as complicated as a wideband power amplifier with feedback. The fact that it regulates the output voltage to a fixed value is incidental. It must be very carefully tested before it is trusted to do its job. The output noise should not be greater than the input stage noise or the noise on the reference voltage. Hum in a regulated supply indicates that it is improperly designed or malfunctioning.

2.5   THE ELECTROSTATIC SHIELD

Sensitive circuits are generally placed inside a metal enclosure. This construction terminates any external electric fields on the outside of the enclosure. The enclosure has capacitances to every part of the circuit and can couple input and output signals together. This coupling is avoided when the enclosure is connected to the reference potential of the circuit.

When one circuit conductor leaves the confines of the enclosure and connects to an external circuit, interference current can flow in this lead. This is illustrated in Figure 2.7. The external field can be represented by a capacitor from an earth point to the enclosure. Current flows in path ① ② ③ ④. This can only be avoided if the enclosure surrounds this conductor and the enclosure is connected to the circuit conductor at its external contact point. This arrangement is shown in Figure 2.8. Note that the interference noise flows in a shield not in a signal conductor. This is the ideal solution and is not always practiced. If the signals are high-level, the interference can often be neglected. If the signals are in the millivolt range, then even 1 mA

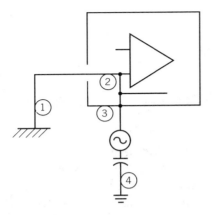

FIGURE 2.7 Interference flowing in a circuit conductor.

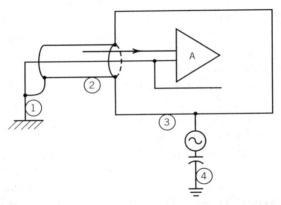

FIGURE 2.8 A proper connection for an electrostatic shield.

flowing in 1 Ω of signal lead is 1 mV. A 1-mA noise current is a typical level found in a laboratory environment.

The exercise given above illustrates four very basic rules in low-frequency electrostatic shielding.

1. A circuit can be guarded from external electric fields by placing it in a metal enclosure.
2. This enclosure should be connected to the circuit reference conductor to avoid complicating feedback paths.
3. If the circuit reference conductor exits the enclosure, the enclosure must include this conductor.

4. The enclosure can only be connected to the reference conductor once, and that is at the point where the reference conductor connects to an external circuit or ground.

2.6 TRANSFORMERS AND THE SINGLE ELECTROSTATIC SHIELD

Transformers are used in most electronic equipment to provide acceptable voltages to operate power supplies and to provide some form of isolation. They are the least accessible component in a design. Once a transformer has been selected, there is little chance for user experimentation. A power transformer provides a capacitive path for interference to enter a circuit and it has a magnetic field that can interfere with circuit performance. Shields can be added, but unless the basic electric field mechanisms are understood, the additions may do little good. The magnetic field problems must be considered separately. The presence of voltages on the coils of a transformer implies that there are electric fields. These fields are needed to make the transformer work, but they also can contaminate signals being handled by the circuit. The electrostatic enclosure described in the previous section is literally breached by the transformer coils. The primary voltages arise from outside the enclosure and the secondary is connected to the reference conductor of the circuit. The coils of the transformer capacitively couple primary voltages to the circuit reference conductor, allowing interference current to flow. The current path is shown in Figure 2.9.

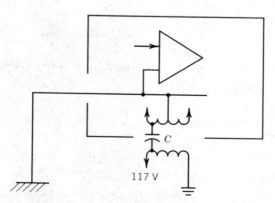

FIGURE 2.9 A power transformer at the enclosure interface.

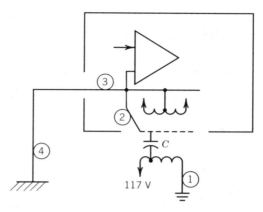

FIGURE 2.10   A single shield in a transformer.

However, a question must be asked—Is there a way to add an electro-static shield in the transformer to eliminate this current flow? Before this question can be answered the internal construction of a trans-former should be considered.

Generally, layers of wire are placed on a bobbin to form the primary coil. The voltage on the outer layer defines the voltage in series with any parasitic capacitances. If the outer layer is at 120 V, then this volt-age is in series with the secondary coil. If there is an interposing shield as in Figure 2.10, this shield is in series with the shield-to-coil capacitance. In this case the primary voltage causes current flow in the input shield.

A secondary coil is constructed so that the first layer of wire is next to the primary coil. This layer is usually at one of the secondary voltages. This voltage can cause current to flow in the parasitic capaci-tances associated with the secondary. If the adjacent conductor is a shield, these currents will flow in the circuit defined by the shield con-nection. In the circuit shown in Figure 2.10, this current flows in path ① ② ③ ④ and this path involves a signal conductor.

The coils of a transformer carry the utility voltage plus any tran-sients or fluctuations present on the line. If the load in the device de-mands pulses of current, the transformer coils will reflect a transient voltage. These voltage transients have high-frequency content and can circulate proportionally more current in the parasitic capacitances of the transformer.

There are four possible connections for a single shield in a trans-former. These are equipment or safety ground, the grounded side of the power source, the local signal reference ground, or power common in the circuit and the enclosure surrounding the equipment.

1. If the shield is tied to safety ground, the secondary voltage forces parasitic current to flow in a signal conductor. The potential difference between equipment ground and signal ground also adds to this current flow.
2. If the shield is tied to the grounded power conductor the problems discussed above are repeated. If the power connection is reversed, the shield is at the power potential and can be dangerous.
3. If the shield is connected to the local signal reference in the circuit, the primary voltage can circulate current in the signal conductor.
4. If the shield is tied to the enclosure, the secondary voltages can circulate current in the signal conductor.

This current flow is not serious when high-level signals are involved. If the lines are short, this current will probably not cause noticeable interference. If this current is to be controlled, then one shield is inadequate. The best connection for a single shield is to equipment ground because a primary short to the shield will trip a branch circuit breaker. Also, if there is a lightning surge it will follow the equipment ground path rather than enter the circuit.

2.7   THE GENERAL ELECTROSTATIC SHIELDING PROBLEM

Transformer shielding for electrostatic isolation illustrates how an interposing piece of metal can be used to control the path for an interfering current. The complete solution to the problem involves providing a known path for parasitic current flow for every interfering potential difference. The correct solution involves three shields as there are three sources of difficulty. These sources are the primary coil, the secondary coil, and the general electrostatic enclosure. By clever transformer construction the number of actual shields can be reduced. In difficult applications the shields may have to be present and there can be no compromise.

Up until this time the discussion has been idealized by assuming the shields are perfect. In practice there are mutual capacitances that

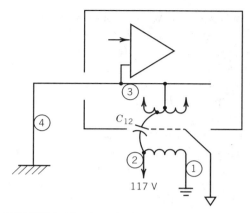

FIGURE 2.11   A typical mutual capacitance.

allow current to flow around shields. Using circuit symbology, Figure 2.11 shows a simple shield and its leakage capacitance C_{12}. The
primary voltage causes current to flow in loop ① ② ③ ④. In 60-Hz
transformers of about 5 W a simple shield consisting of a layer of copper or aluminum might have a leakage capacitance of 5 pF. Without
the shield the parasitic current flow might be 30 times greater. Shields
that literally box in the primary coil can be constructed so that the
leakage capacitance is held to below 0.1 pF. These shields are expensive to execute and they limit the performance of the transformer.
These very small capacitances can easily be violated by improper wiring techniques. For example, the primary leads may have to be
shielded within the equipment so that the mutual capacitances to the
primary leads can be held to below 0.1 pF. If this is not done, the fine
shielding in the transformer is wasted.

The signals that cause interference are not limited to the power fundamental. Interference is often high-frequency in character and there
are limits to how carefully a transformer can be built and installed.
There are isolation transformers on the market that claim mutual capacitances below 0.01 pF. The more important claim should be related
to how much interference current flows for each user configuration.

There are subtleties to the shielding problem. For example, a shield
is a single turn and cannot short out at the overlap. This turn limits
the shields effectivity as the transformer voltages appear on the shield.
A shield tie, if brought out, should not couple to the leakage flux of
the transformer or this adds to external current flow. Both of these

mechanisms allow the shield to be violated and limit the effectiveness of the shield. The use of the multishielded isolation transformer at high frequencies and in large systems is discussed later.

In certain transformer types shielding is not possible. This includes the autotransformer that provides step-up or step-down voltages by tapping the one primary coil. Since this type of transformer uses the grounded power conductor or neutral as one of its connections, it is illegal to reground this point to any earth or other grounded conductor. To do so would violate all the safety provisions of the code. If the autotransformer is mounted and used inside of hardware after another suitable transformer, then the grounding restrictions may no longer pertain.

2.8 MAGNETIZING CURRENT

Lenz's law relates the *B* field in the iron to the rate of changing voltages in a coil surrounding the field. The magnetizing current required to establish this induction field is determined by the permeability of the iron. As the *B* field increases, the iron tends to saturate and the permeability falls. In effect this means that the magnetizing current must increase. In designs that operate with a poorer grade of iron or with a high *B* field, this magnetizing current can be a source of interference.

The induction field in the iron increases as long as the voltage is positive. At the voltage zero crossing the voltage reverses sign and the *B* field starts to decrease. At the next zero crossing of voltage the *B* field is at its other extreme. The highest magnetizing current required will flow just before the voltage reaches a zero crossing.

The magnetizing current flows in the leakage inductance of the transformer. In practical terms this means that the magnetic fields near a transformer increase during this point in the cycle. This increase in current also flows in the primary conductors and appears as a magnetic field proportional to the primary loop area.

2.9 IMPEDANCE CONVERSIONS

A step-down transformer reduces the source impedance to loads by the turns ratio squared. This is important for capacitor input-type rectifier loads where the demands on peak current are significant. A low impedance keeps the voltage waveforms from being corrupted. In systems where the waveforms are badly distorted this can be an interference problem for other loads. Transformers can also function as passive filters to reduce this effect.

2.10 COMMON-MODE AND DIFFERENTIAL-MODE SIGNALS

Interference often couples to a group of conductors such as an entire power cable. This form of coupling is said to be common mode in nature. Strictly speaking, a common-mode signal is an average signal with respect to a reference conductor. The reference conductor is usually taken as a ground or signal reference conductor. The term longitudinal mode is used in the telephone industry instead of common mode.

In a single-phase power connection the average voltage is approximately 55 V referenced to earth ground. This voltage is not interference and it is not the voltage that appears next to a shield in a power transformer. It is thus incorrect to say that all common-mode voltages are bad and should be rejected.

The term normal-mode signal means the difference signal. In most cases the difference signal is the signal of interest. In many applications, interference can take the form of a difference signal. Unless this signal is out-of-band, there is no way to reject it. The telephone industry uses the term transverse mode instead of normal or differential mode.

2.11 ISOLATION TRANSFORMERS

A variety of shielded transformers are available on the market that are called isolation transformers. The user is usually left with the problem of connecting the shields to reduce interference. The mutual capacitances are often specified to be less than 0.1 pF. If the windings and shields are not very carefully routed, this performance is easily violated. To maintain this specification the primary leads should be kept in a shield.

In many applications the intent is to use the transformers to reduce high-frequency interference and this is where the mutual capacitances are critical. If the shield lengths are long, the inductance of these paths makes the shields ineffective. If the primary and secondary leads are routed in the same conduit or in the same receptacle, the shielding is violated.

Transformer shielding becomes ineffective above about 30 kHz. The usual practice is to incorporate passive filters on the power line to limit high-frequency interference coupling. The transformer by its very nature is a low-pass filter for high-frequency differential interference. The mutual capacitances limit the performance for common-mode rejection.

If the transformers are used outside of equipment, then all of the safety aspects of the National Electrical Code must be considered. The primary and secondary conduit, along with the transformer case and core, are all classed as equipment ground and must provide a low-impedance path for any possible fault condition. The conduit may not be isolated or insulated in any way.

2.12   COMMON-MODE AND DIFFERENTIAL-MODE SHIELDS

The single shield in a power transformer can be used to reject power-line common-mode interference. The shield is tied to the local equipment ground or local signal reference conductor. In Figure 2.12 the common-mode source V causes current to flow in path ① ② ③ ④. The mutual capacitance C_{25} allows some current to follow the path ① ② ⑤ ⑥. The shield can be tied internally to the case and core of the transformer. If this shield is made available for user definition, it should never be tied to a separate earth ground or run any distance. A separate earth ground could bring a large potential difference into the transformer during lightning activity and blow up the transformer and other equipment.

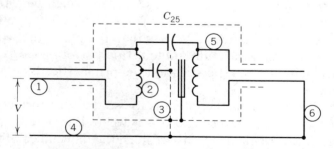

FIGURE 2.12   A common-mode shield.

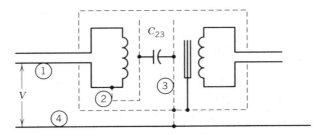

FIGURE 2.13 A power isolation transformer with two shields.

Common-mode current in Figure 2.12 flows in turns of the primary coil. This current flow appears in the secondary load by transformer action. This conversion of interference from common to differential mode can be eliminated by adding a primary shield as in Figure 2.13. The common-mode current now flows in path ① ② ③ ④ and avoids the primary turns. This shield can be tied internally by the manufacturer or locally by the user. This shield is said to "resist" differential-mode conversion.

If the load is electrically noisy then a secondary shield can be used to control the flow of interference from the load to the line. This type of problem should be treated in conjunction with a passive filter located on the secondary or load side of the transformer. This shield should be connected to the grounded side of the secondary, which may be a centertap. In some commercially available transformers the connection is made internally.

2.13 DISTRIBUTION TRANSFORMERS

Distribution transformers in a facility help to isolate loads and improve load regulation. They are classed as a separately derived power source and as such must be treated like a new service entrance. The Code requires that the secondary be grounded to the nearest point on the grounding electrode system. This is the only grounding of the secondary power conductors that is allowed. All equipment grounds are returned to the panel associated with this new power source to provide a low-impedance fault-protection path.

It serves very little purpose to provide shields in this type of transformer unless the transformer is mounted very near and preferably on the ground plane associated with the equipment being powered. The same rule applies to line filters. It is preferred to mount line filtering at or near the load requiring protection. Line filters are treated later in this book.

Computer power centers are available that provide shielded transformers, breakers, filtering, and surge suppression in one unit. These units are designed for mounting near the loads being served.

2.14   MAGNETIC SHIELDS

Low-frequency magnetic fields are not easily contained. It is usually wise to limit the problem by reducing the coupling to the circuit in question. This involves reducing the coupling loop area to the minimum. When this is not practical, the magnetic shield can be applied.

The fields near a transformer core are determined by the flux level in the iron. The biggest problem is the leakage flux external to the transformer that is developed by the load current. For small transformers the case can be made from nested steel and copper cans. The size of the conductor opening must be limited. The inner can should not be an exotic material as it can easily be saturated. These nested configurations can be quite expensive and are generally avoided.

Shading coils are often wrapped around a transformer core to reduce external fields. These coils act as shorted turns to the leakage flux and tend to redirect and contain the external field. These turns must carry a lot of current and are usually large copper conductors.

The energy stored in a magnetic field is reduced when there is a better magnetic path provided. Nature always finds a field configuration that stores the least amount of energy. This means that magnetic materials can be used to redirect the field away from a sensitive area. A steel plate near a connector can be used to carry the flux away from the connector so that it does not couple to the pin connection area inside the connector.

2.15   UNGROUNDED AC POWER SOURCES

There are places where ungrounded power is acceptable or the norm. The Navy uses ungrounded power on shipboard to avoid electrolysis and to make the system more fault tolerant. The first fault, for example, will not trip a breaker.

Ungrounded power is the norm in parts of Europe. The conduit used to carry these conductors is grounded to maintain a shock-free system. In industrial applications where a fault and subsequent loss of power could do a lot of damage, floating systems are acceptable. An example of this is an electric furnace where the melt must remain molten. These ungrounded or floating systems tend to be noisier electrically than grounded systems. Every time a load is switched, a common-mode voltage is developed. Power for computer equipment delivered to Europe must often be conditioned to limit this type of interference.

2.16 UNGROUNDED DC POWER SUPPLIES

Floating dc power sources are used to excite transducers or approximate the performance of a floating battery. The transformer that supplies this power still capacitively couples the secondary circuit to the primary circuit. The quality of the circuit is measured by the current flow in a grounding resistor, as shown in Figure 2.14. If the current is to be held to less than 10 nA at 60 Hz, the leakage capacitances C_{12} and C_{34} must be below 0.22 pF. The three shields are required and the leakage capacitances associated with each shield can contribute to the current flow.

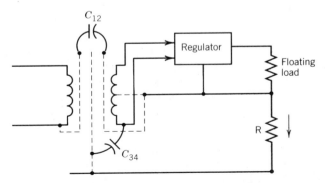

FIGURE 2.14 A floating power supply.

2.17 SWITCHED-MODE POWER SUPPLIES

Switching transistors can be used to convert dc to a high-frequency square-wave voltage usually below 100 kHz. This square-wave voltage is then applied to a small transformer that conditions the voltage to new levels for rectification. In this manner 150 V dc can be converted to standard voltages such as 5 V or ±15 V. In some circuits the switching duty cycle is varied to regulate the resulting dc voltage.

Power can be taken directly from the power line to avoid a big transformer. When the input voltage goes to zero there is no power available, so energy storage in the form of an electrolytic capacitor is usually supplied. The switches are either fully on or fully off so that they dissipate a minimum amount of heat. This type of power supply is cost effective and is appearing in more and more circuits. These converters allow systems to be designed that work at any power frequency including dc without making circuit changes. Even the fans are operated at a fixed ac voltage after the energy has been converted from ac to dc and back to ac. One impact on the power system is the extensive use of capacitor input filters with its associated large peak current demand.

A serious problem posed by switching regulators is the flow of high-frequency switching current in the grounding system. If the line filters and power-supply filters are referenced to various grounds, significant amounts of interference current can flow. This current has content at

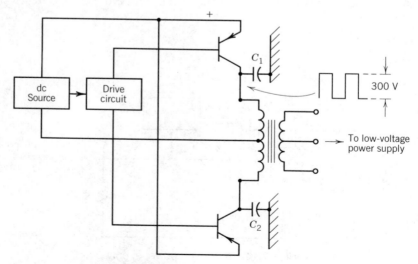

FIGURE 2.15   A switching regulator.

the harmonics of the switching frequency and this content is not easily isolated by shielding. The problem has become serious enough that various regulatory agencies limit the capacitance and current flow to equipment ground.

Although they are very efficient, switching elements must still dissipate heat. They are usually mounted to equipment ground as this is the best available thermal radiator. The reactive current flowing in the capacitance from the collector or drain of the transistor to equipment ground can be one of the major sources of interference. This current flows in C_1 and C_2 as illustrated in Figure 2.15. The voltage on the collector swings 300 V at each switch transition. Assume the switching rate is 50 kHz and the rise time is 0.2 μs. The current in 30 pF for this rise time is 45 mA. If many parallel switching elements are involved, the current can literally be amperes.

The insulator between the transistor and the equipment ground can be supplied with a conducting shield.* When this shield is connected to the circuit reference conductor, the currents circulating in the equipment ground can be reduced by an order of magnitude. This circuit is shown in Figure 2.16. The largest fraction of the switching current follows the path ① ② ③ ④.

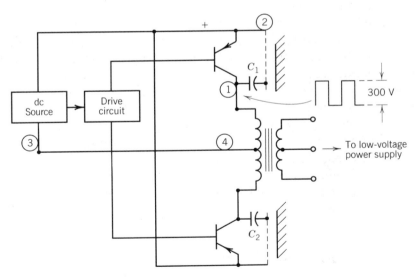

FIGURE 2.16   A shielded insulator for the switching transistors.

*Sil Pads are manufactured by Berquist of Minneapolis, MN.

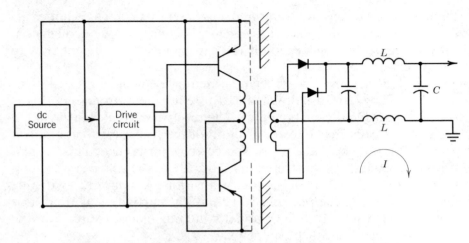

FIGURE 2.17 Load filters to limit common-mode current flow.

It is usually impractical to build a switching transformer with internal shields. A shield would increase the winding capacitances and the leakage inductance. At the frequencies involved the mutual capacitances would limit the effectivity of the shield. Most of these transformers are toroids, which makes the mechanics of shielding very difficult.

Switching transformers use ferrite-type cores, which have, permeability at the switching frequencies. The few turns required limit the capacitance between windings. The fast-changing voltages appear between windings and can cause significant reactive current to flow in the grounding system. The filter in Figure 2.17 shows how this common-mode leakage current can be further controlled. The series inductors must have a high enough natural frequency to be effective.

The common-mode current flow in a switching-power supply can be measured by inserting a small impedance in series with the secondary grounding connection. The voltage across this impedance measures the current flow. If the secondary circuit is floating, the filter will serve no purpose.

2.18   THE MOUNTING OF TRANSFORMERS

The laminations used in small transformers are often provided with a round hole that can be used to mount the transformer. Metal screws should not be used to mount the transformer onto a conductive surface. This geometry allows magnetic flux to couple to the loop formed by the screw and the metal surface. This coupling causes current to flow in this loop, which is reflected as added primary current. This current can cause potential drops in the metal which can couple to the circuitry.

Large-distribution transformers are often mounted to a steel framework. The flux from these transformers can couple to any loop of metal in the vicinity of the transformer. This coupling can result in significant power current circulating in the building steel. Because there are so many parallel paths in a typical steel framework, the current flows in all parts of a building. The initial coupling can be avoided by breaking all nearby loops of metal. This can be done by adding insulators and insulating washers at appropriate points in the mounting metalwork. It is not uncommon for currents of 10A to circulate in steel members hundreds of feet away from the offending transformer. This level of current can be very surprising in a facility that is properly wired per the National Electrical Code.

TRANSMISSION LINES IN ANALOG AND DIGITAL TRANSMISSION

3.1 ELECTRICAL TRANSPORT

All electrical energy is transported by fields. This includes power at dc, power from the generators at a hydroelectric station, and power from a radar antenna. Energy at low frequencies is more easily transported between conductors than in free space. As an example, the majority of the energy from a power station goes where the conductors lead the fields. At 600 Hz enough field energy is lost per mile to make this frequency unacceptable.

When radiated energy enters a system of conductors, the fields tend to follow the conductors in a manner roughly based on the ratio of impedance presented by the conductors to the impedance of free space. This is Nature's way of taking the easiest path. Some of the energy couples to the conductors and literally takes a ride wherever the conductors go. In most cases the conductors are not well-designed pathways and the energy is lost, reflected, or reradiated.

Field energy can be transported between any conductors, including conduit, shields, earth, power grounds, equipment grounds, input leads, output leads, and control leads. All conductor pairs are two-way streets; that is, energy can flow in either direction. Conductors can carry field energy into or out of a system. Nature pays no heed to color codes, labels, intended use, or signals already present.

A closed metal box made from a perfect conductor allows no electromagnetic energy to enter or leave the box. This isolation is violated if

45

one conductor enters through a tiny hole. This conductor can be a shield or a grounded conductor. In theory, every penetrating lead must be filtered to the box at the surface. Bringing the lead inside the box before it is filtered violates the filter. The field level in the box depends on the nature of the external field. The sensitivity inside the box is determined by the type of equipment installed in the box.

3.2 AN IDEAL TRANSMISSION LINE

A pair of conductors can be called a transmission line. The ideal qualities of these lines are well understood and form the basis for understanding the transport of interference once it has coupled into the conductors. The ideal case is a good place to start to appreciate the phenomenon.

Consider two long parallel conductors connected to a switch and a battery as in Figure 3.1. At the moment the switch is closed there is no field energy stored in the transmission line. Energy transport in zero time requires infinite power. This means that it takes a finite time for the fields to manifest themselves. Fields cannot move any faster than the speed of light, which is about a foot every nanosecond. In a typical transmission line the speed is approximately half this value.

Electrical field energy must exist wherever there is a potential difference. The capacitance per unit length requires that a charge must be stored if there is a voltage. This charge must be carried by a current that flows in the conductors. This current implies energy storage in a magnetic field. As the wave progresses down the line, both an electric field and a magnetic field store energy behind the wavefront. This means that in the ideal case the current flow is uniform and steady. A steady current and a steady voltage represents power into a resistor. This resistance value describes the characteristic impedance of the

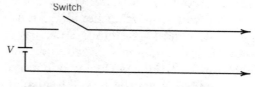

FIGURE 3.1 An ideal transmission line.

line. The word impedance is used because the capacitance and inductance are not ideal.

3.3 THE FIELD PATTERNS IN A TRANSMISSION LINE

The electric and magnetic field patterns behind the wave front are shown in Figure 3.2a. The field energy is confined to the region between the two conductors, and in both cases the field vectors are perpendicular to the direction of energy flow and perpendicular to each other. This is a basic fact in all electrical energy transfer. The E field and the H field are at right angles to each other and to the direction of power transfer, which is at right angles to both fields. This power vector is called Poynting's vector, which is a point concept and represents the power per unit area at a point in space. A surface integral is required to total all the energy crossing a surface perpendicular to the transmission line.

The field pattern between a round conductor and a conducting plane is the same as the field pattern in Figure 3.2a except that the lower half of the field is missing. The same field intensities require that the voltage from the conductor to the plane be one half the voltage in Figure 3.2a. This means that the characteristic impedance of this configuration is one half. This field pattern is shown in Figure 3.2b. It is important to note that the fields are confined to the region between the conductor and the conducting plane. The current flow in the plane is confined to the area where the field lines terminate. If the conductor is brought close to the conducting plane, the field is more tightly confined.

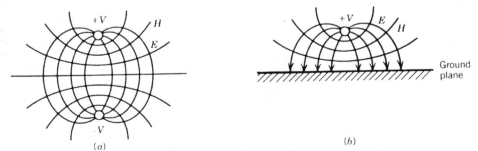

FIGURE 3.2 The field pattern along a transmission line.

3.4 TRANSMISSION LINE TERMINATIONS

When the transmission line in Figure 3.2 is terminated in its characteristic impedance the voltage–current relationship at the terminating impedance is equal to that of the line. As far as the source is concerned, the line looks infinite in length. Current continues to flow after the wave reaches the end of the line and now energy is dissipated rather than stored.

When a transmission line is left unterminated, a voltage reaching this point reflects. The return wave must return the arriving energy and at the same time cancel the current at the end of the transmission line. The reflected wave must have the same voltage and opposite current. The voltage at the termination end is the sum of the two waves and will be double the voltage at the send end.

When a transmission line is terminated in a short, the energy at the termination is also reflected. This time the reflected plus the initial wave must cancel the voltage. The sum of the two waves is double the current at no voltage. When this reflected wave reaches the source again, the third wave must be transmitted that reestablishes the voltage. At this time the sum of the three waves has tripled the current. After many reflections the current has increased to the point where the short circuit is felt by the source.

Step waves traveling back and forth produce steps in current flow and this is not compatible with the concepts of impedance generally applied to circuits. Impedance is correctly a sinusoidal concept where the phase and amplitude of the resulting current measures performance. The ideal transmission line terminated in its characteristic impedance looks like this impedance at all sinusoidal frequencies. When the terminating impedance is anything else, the source impedance depends on termination impedance, line length, and characteristic impedance.

When a transmission line is terminated in a short, the reflected sinusoidal wave can arrive back at any phase angle. If the reflected wave returns back in phase with the initial signal the input impedance is infinite. If the reflected wave returns back with a phase reversal, the input impedance is a short. If the transmission line is the right length, the input impedance can be pure reactive like a capacitor or inductor. In the case of noise phenomena, the parallel geometries that are encountered are rarely terminated and the lengths are uncontrolled. This makes an accurate analysis essentially impossible.

The impedance presented by a transmission line to signals that cover a wide spectrum is complex. An analysis would require that the spectrum of the incoming signal be first established. The response at each frequency would then have to be calculated. The final result would be the composite of all the responses. This is difficult enough, but when the energy is captured by these poorly defined geometries the best way to describe the result is to say that the energy sloshes back and forth until it is dissipated. With a few simplifying assumptions the magnitude of the worst case scenario can be calculated.

3.5 THE COAXIAL CABLE

The fields from a parallel wire transmission line are very nearly confined to the area around the conductors. Some of the field energy does leave the confines of the circuit and radiate. To completely confine the fields, one of the conductors can be wrapped around the other. This geometry forms a coaxial transmission line. The quality of the cable is determined by the type of dielectric and the nature of the sheath. This outer sheath can be a simple braid or a smooth, carefully made tubing. These design differences define the qualities required for high-frequency transmission. These differences also relate to how easy it is for field energy to couple into the cable or for energy to escape from the cable.

The parallel conductor transmission line required that the current flow was limited to the two conductors. This same constraint must hold for coaxial cable. The fields are contained only when the current in the center conductor is returned on the outer sheath. If any current is allowed to return on another path, the field is outside of the cable and the cable is not used as coax. In practice the majority of the current will actually flow on the inside surface of the sheath as this is where the field terminates. If the energy is confined by the sheath, the cable need not be grounded or terminated to function as coax.

Coaxial conductors with the shield connected at one end only are not being used as coax. The outer sheath is simply used as an electrostatic shield. Single-point shield ties are needed at low frequencies. This means that expensive coaxial cable is not needed for low-frequency shielding.

When a coaxial cable is grounded at both ends, there are obviously other ground return paths for the signal. The question most often asked is why doesn't current return over this other path? The fields involved tell the story—current cannot take an alternate path unless there is field external to the cable.

3.6 WHAT FREQUENCY TO SELECT FOR ANALYSIS WHEN THERE ARE NO SINUSOIDS

The proper way to analyze a problem involving pulses or nonsinusoidal waveforms is to define the spectrum of sinusoids that makes up the signal. Pass each sinusoid through the system and then recombine the resultant sine waves to obtain the end result. This is difficult and a first-order simplification is needed.

Most electromagnetic coupling is area related. Coupling for sinusoids increases linearly with frequency. For most transient phenomena, the frequency content falls off linearly with frequency out to a frequency defined by $1/\pi\tau_r$ where τ_r is the rise time. These two facts imply that most of the spectrum in a transient event couples so as to reproduce the initial pulse, but limited to an upper frequency of $1/\pi\tau_r$.

Therefore, it makes sense to solve the problem at this one highest frequency and accept the amplitude of the response as representative of the amplitude of the actual pulse that is coupled. For most practical applications this analysis will be accurate to within 25%, and this is accurate enough to predict whether a problem exists.

3.7 CROSS COUPLING

The fields established by a signal often share the same space used by a second signal. This sharing of space results in crosstalk. If the fields are confined to specific areas such as inside sections of coax, the coupling is essentially zero. Fields can enter or exit coax through the sheath, but these fields are usually highly attenuated.

Coupling between unshielded cables can be differential or common mode in nature. The conductors can be pairs of wires or wires over ground planes (large conducting surfaces). The loop areas considered determine the nature of the coupling. If the loop area involves the signal path, the coupling is differential. If the loop area involves a reference plane, the coupling is common mode. The interfering field can result from common- or differential-mode signals.

Coupling between unshielded cables involves distributed capacitances and inductances. A circuit theory approach to these problems can be rather mathematical. Useful results can be obtained by using lumped parameter models rather than the more complex distributed parameter models. The procedure involves solving the problem first as a simple capacitive divider and then as a two-coil mutual inductance. One of the mechanisms will usually dominate and then the other can be discarded. If both coupling mechanisms are equal, they are considered additive.

The capacitive coupling problem involves the voltage divider action between the source or culprit voltage and the receiving or victim impedance. The source impedance is not considered since the culprit voltage defines the coupling. The inductive coupling is just the opposite. The culprit load impedance defines the current flow, which determines the magnitude of the magnetic field. The voltage induced into the victim circuit is proportional to the area of the victim's loop. In most practical applications the victim's circuit impedance can be neglected.

The half-field pattern of two round conductors has the same intensity as the field pattern of a round conductor over a ground or reference plane. Cable-to-cable coupling is a matter of field intensities. This means that the problem can be first solved as conductors over a ground plane and then corrected for the geometry in question.

The attenuation factors in Table 1 are given in dB. That is, if the source voltage is 1 V, the attenuation factor can be used to calculate the coupled voltage in dB volts. The length is normalized to 1 m and the impedances are normalized to 100 Ω. In the case of capacitive coupling the source and terminating victim impedances in parallel are 50 Ω. If other impedances are involved, the capacitive coupling is proportionately greater if the parallel impedance is greater than 50 Ω. For inductive coupling, the load impedance is compared to 100 Ω as this is the impedance that defines the current flow.

The source and terminating impedances for semiconductors are nonlinear. This means that an average or representative value must be selected. Typical values for TTL logic are 150 Ω.

When the victim and culprit conductor distances are different the value of d should be the geometric mean. The table is only approximate, but it does provide some insight into the magnitude of the coupling. A careful mathematical assessment is very difficult to handle and often is no more meaningful.

Table 1. Normalized Cable-to-Cable Coupling

Signal Pair 1 / Signal Pair 2, Use $d = \sqrt{d_1 d_2}$

Signal 1 / Signal 2, Use $d = \sqrt{d_1 d_2}$

d and S are in millimeters

Inductive Attenuation for Culprit $z = 100\ \Omega$, Length $= 1\ m$

	$d = 0.5\text{–}3$ mm			$d = 3\text{–}30$ mm				$d = 30\text{–}300$ mm			
	$s = 1$	$s = 10$	$s = 100$	$s = 1$	$s = 10$	$s = 100$	$s = 1000$	$s = 1$	$s = 10$	$s = 100$	$s = 1000$
1 kHz	−107	−139	−179	−96	−107	−139	−219	−91	−96	−107	−179
3 kHz	−97	−130	−169	−86	−97	−130	−209	−81	−86	−97	−169
10 kHz	−87	−119	−159	−76	−87	−119	−199	−71	−76	−87	−159
30 kHz	−77	−110	−149	−66	−77	−110	−189	−61	−66	−77	−149
100 kHz	−67	−99	−139	−56	−67	−99	−179	−51	−56	−67	−139
300 kHz	−57	−90	−130	−46	−57	−90	−170	−41	−46	−57	−130
1 MHz	−47	−79	−119	−36	−47	−79	−159	−31	−36	−47	−119
3 MHz	−38	−70	−110	−27	−38	−70	−150	−22	−27	−38	−110
10 MHz	−28	−60	−100	−17	−29	−61	−140	−14	−18	−30	−101
30 MHz	−20	−52	−92	−11	−22	−55	−132	−8	−13	−24	−94
100 MHz	−13	−45	−85	−7	−18	−50	−125	−5	−10	−21	−90

Capacitive Attenuation for Victim z = 50 Ω, Length = 1 m

1 kHz	−104	−140	−180	−220	−103	−122	−155	−195	−103	−117	−130	−163
3 kHz	−95	−131	−171	−211	−93	−113	−145	−185	−93	−107	−121	−153
10 kHz	−84	−120	−160	−200	−83	−102	−135	−175	−83	−97	−110	−143
30 kHz	−75	−111	−151	−191	−73	−93	−125	−165	−73	−87	−101	−133
100 kHz	−64	−100	−140	−180	−63	−82	−115	−155	−63	−77	−90	−123
300 kHz	−55	−91	−131	−171	−54	−73	−105	−145	−53	−67	−81	−113
1 MHz	−44	−81	−120	−166	−43	−62	−95	−135	−43	−57	−70	−103
3 MHz	−35	−71	−111	−151	−34	−53	−86	−125	−33	−47	−61	−93
10 MHz	−26	−61	−101	−141	−24	−43	−75	−115	−24	−37	−51	−83
30 MHz	−19	−53	−93	−133	−17	−34	−66	−106	−17	−29	−41	−74
100 MHz	−14	−46	−86	−126	−12	−20	−58	−98	−12	−20	−33	−65

In using Table 1 the following facts must be remembered:

1. Coupling is proportional to length up to a quarter wavelength.
2. Capacitive coupling rises proportional to victim impedance.
3. Inductive coupling rises as the culprit impedance falls.
4. The conductor spacing s is the geometric mean between the victim and culprit spacing.
5. The cable spacing S is given for three ranges. If necessary the attentuation values may be corrected for different spacings by linear interpolation.
6. The frequency of interest is $1/\pi\tau_r$ where τ_r is the rise time.

3.8   CABLE CONFIGURATIONS

A variety of cable geometries are available to interconnect circuits. Coaxial cable can be used to transport field energy in a transmission line sense. Shielded conductors can be used to carry low-level low-frequency signals. Open conductors are used for the transport of power, high-level signals, and control. Each cable geometry has a different mechanism that dominates interference coupling.

Coaxial cables are too costly and cumbersome to use in the interconnection of many digital circuits. Designers can specify ribbon cables with and without associated ground planes. These cable types come under the trade names of Flexstrip and Microstrip. Ribbon cables are also available with certain conductor pairs twisted to reduce differential field coupling, radiation, or cross coupling between conductor pairs.

Interference and radiation from various conductor geometries is a complex subject. One of the most critical areas is cable termination. The bonding or treatment of the sheath at both ends of a cable run usually defines the performance not the quality of the cable. On long runs the cable quality does become more important. A review of each cable type and the coupling problems will help to clarify this point.

3.8.1. Open Conductor Pairs

Coupling is a function of proximity and loop area. If the victim loop area is reduced, the coupling is reduced. If the dominant mode is inductive, the loop area of the culprit can be reduced. Twisting the culprit leads can also be beneficial. If a ground plane is involved, routing the conductors on the plane will reduce the coupling.

3.8.2. Shields at Low Frequencies

Low-level signals are often transported using two-wire twisted shielded cable. Multiple grounding of this shield is improper as the shield cannot "short out" grounds. Multiple grounds will only establish a potential gradient along the shield. If one end of the shield is at the zero-signal reference potential, the remaining shield will be at some interfering potential. This will couple capacitively to the signal and add noise or interference. The correct use of a shield at low frequencies is to connect the shield once to the signal common at the point where the signal grounds. This is very important for low-level input signals and relatively unimportant for high-level low-impedance signals. If there is a high-frequency interference, then a secondary solution must be superposed on the correct input shield treatment. An example of this treatment might be running many twisted-pair shielded cables in one metal conduit that can be multiply grounded.

3.8.3. Coaxial Shields

The function of the sheath on coaxial cable is to confine the signal field to the inside surface of the sheath. At the termination of the cable, the fields are not as easily contained. The impedance of the sheath itself can be milliohms at 10 MHz. At this frequency the impedance of a small loop of wire forming a pigtail connection to the sheath can look like a few ohms. If a pigtail is used, the sheath is not adequately terminated. This means that fields may not follow the coaxial path and that interference can enter the cable at this junction.

1. Shield terminations should not enter the equipment through a connector. Noise current on the shield can radiate into the hardware.
2. The shield connection should be to the outside of a bulkhead and should not be formed into a half turn.
3. If practical, the shield itself should be bonded to the connector at more than one point. The bonding surface for the connector cannot be painted, anodized, or subject to corrosion.

4. In critical applications the shield should be terminated in a backshell connector. This connector must be properly bonded to the bulkhead on a prepared surface.

3.8.4. Shields and High-Frequency Problems

Cables with aluminum foil shields are in common use. The foil is fragile and difficult to terminate. Manufacturers often run a drain wire on the inside of the shield to make contact with the shield. The termination of the actual shield is often nonexistent and noise current can easily follow the drain conductor which is bonded to the equipment bulkhead or connector shell. This noise couples to other conductors in the cable and at high frequencies this can be very unsatisfactory. A foil bond can be easily destroyed by mechanical motion of the cable. A drain wire on the outside of the foil is preferred over an inside drain.

Braided cable is used in many low- and high-frequency applications. At low frequencies the optical coverage defines the shielding against external electric fields. At frequencies above 100 kHz, the braid and skin effect play a role in keeping field phenomena from penetrating the sheath. Some of the currents on the outside of a braided sheath can follow the weave and enter the inside surface of the cable. Currents on the inside surface implies field on the inside of the cable. This mechanism allows interference to enter a cable. Finer braid or double braid helps to reduce this coupling. The best solution is to use a solid thin-wall tubing as a shield and this avoids the weaving that allows fields to penetrate. Of course if the thinwall is not terminated correctly on the ends of the cable, the thinwall will be ineffective.

3.8.5. Ribbon Cable

Digital transmission on ribbon-cable conductors involves a compromise which separates signal and signal return conductors. A common mistake is to use only one return or ground conductor for a group of signals. The correct approach is to provide a return path adjacent to every signal. This preferred arrangement is shown in Figure 3.3. These ground conductors are in effect an extension of the ground plane used on the printed circuit board. This means that every ground conductor must be bonded to the ground plane on each end of the ribbon-cable run. If a ground plane is provided with the ribbon cable, it must be terminated along its entire width to the bulkhead or circuit ground

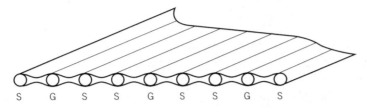

FIGURE 3.3 Ground conductors in a ribbon cable.

plane at each end. If this is not done, the qualities expected of the improved cable will not be realized. The cable will more easily radiate or
be susceptible to external fields.

3.9 IMPEDANCE PER SQUARE

The resistance of any conductor is equal to the resistivity ρ times the
cross-sectional area A times the length l or

$$R = \rho l/A \qquad (1)$$

If the area is a square equal to $l \times t$ where t is the thickness, the resistance is

$$R = \rho/t \qquad (2)$$

This means that the resistance of a square of metal is independent of
the size of the square and depends only on thickness. This is why a
thin metal surface is characterized by the measure ohms-per-square.

A copper sheet 1 mm thick has a resistance of 17.2 $\mu\Omega$ per square.
This means that if a sheet of copper 1 mm thick and 20 ft square carries a uniform current of 1000 A, the voltage drop would be 17.2 mV.
At 1 MHz the impedance rises to 36.9 $\mu\Omega$. At this frequency a uniform 1000-A flow would cause a voltage drop of 36.9 mV.

3.10 POINT CONTACTS ON METAL SURFACES

The very low impedances described for sheets of copper require that
the current flow utilize the entire sheet. When typical point contacts
are made, the impedance rises by about 30%. The impedance of a

typical circuit involving finite lead lengths is dominated by the leads, not the copper sheet. For example, if the leads are 100 in of #10 wire at a frequency of 1 MHz, the impedance rises to 62.8 Ω. It does little good to make this conductor size #0000.

The lesson is obvious. When current flow must concentrate, there is inductance and this implies an impedance. When current is dispersed, there is little field energy storage and the impedance is low. This is the basic quality of a ground plane. Surfaces made with 1 mm of copper are rather fragile and therefore are not used. Thicker sheets of aluminum or steel are almost as good and are quite adequate conductive planes.

3.11 GROUND PLANE

A conducting surface is a low impedance for surface current flow even at 10 MHz. The potential drop on the surface is below millivolts in most practical applications. In effect, a ground plane comes very close to being an equipotential plane. This one property makes a ground plane useful. There is essentially no tangential E field along this surface.

A ground plane does not stop electromagnetic radiation any more than the earth stops radio transmission. Energy can propagate along the surface of a ground plane as long as there is no tangential E field. Energy that arrives at the surface with a tangential component of E is simply reflected. This means that there must be an H field at the surface and this implies that there is surface current. This current is flowing in a low impedance and does not represent a sizable potential drop. The term plane is misleading. A ground plane can bend and follow a curved surface. A ground plane can interconnect two floors in a facility or follow a cable in a tray.

3.12 THE CONCEPT OF VOLTAGE

In Section 1.21 voltage was defined as the work required to move a unit charge in a field. When a voltmeter is used to measure a voltage, the unit charge must follow the voltmeter leads. If the leads couple to a changing field, the voltmeter measures the field strength as well as the potential drop between the points of measurement. To avoid this coupling the voltmeter leads must be dressed along the ground plane. This eliminates coupling to spurious fields.

Engineers will often measure a voltage between two grounded points in a facility. The voltmeter or oscilloscope leads cannot follow the steel girders in the building and thus the leads couple to fields in the room. If an attempt is made to short these two ground points together with a large copper conductor, it will not work. The fields that are present will persist even with the added copper. In truth, if the steel of the building is not adequate, the added copper can also do very little. Even at 1 MHz a run of 10 ft of #4 wire looks like many ohms. If the #4 wire is pounded out flat to cover a large surface area, the equipotential plane so formed could allow a low-voltage measure to be made between points in the facility. The voltmeter leads would have to route along the plane. If a voltmeter can be routed to avoid field coupling, then any signal cable can be handled in the same way.

3.13   GROUND PLANES IN PRINTED WIRING BOARDS

The ground planes used in digital circuits are found mainly on printed wiring boards. The ground plane serves as the signal return path and power return path for every logic circuit. Every signal path and the ground plane form a transmission line. The return current path for each signal is confined to an area under the trace. This means that the field for each signal is confined to a small region. Fields further away from adjacent signals do not share the same space. This geometry limits crosstalk, radiation, and coupling to external fields. Ground planes make it possible to operate digital circuits at high clock rates.

When a logic element changes state, the logic signal must propagate down a transmission line. The energy supplied to the transmission line must come from a power supply. If this energy is not immediately available, the energy sent forward must be robbed from the power-supply transmission line. If the lines all have equal characteristic impedances, the voltage sent forward would be one half the supply voltage. This is the most undesirable condition for a logic signal. Energy for each signal transmission must come from energy storage on the board. This energy can be supplied from local decoupling capacitors. The physical area involved in this path defines an inductance which appears in series with the decoupling capacitor. The capacitor itself has inductance which involves leads and foil lengths. If several logic elements share this one energy source, the fields must share the same physical volume. As logic speeds increase, this type of crosstalk becomes critical.

A common error in providing decoupling capacitors is to make them too large. Smaller capacitances have higher natural frequencies and can supply their energy more rapidly. If the energy is not available locally or it is not available in a short time, the system is apt to radiate and appear noisy.

Radiation from a printed circuit board is directly related to loop areas. These loop areas arise from the connection geometries to each integrated circuit and the decoupling capacitors. Multiple ground planes do not limit these areas and therefore do not control the radiation. Surface-mounted components reduce loop areas and therefore reduce both radiation and susceptibility.

A power plane is basically no different than a ground plane except that it supplies a power connection. Signals that are transported on traces between a ground and power plane may see a different characteristic impedance and the field geometries may be more closely confined. Radiation is still a function of component geometries. The parallel ground planes act as a low-impedance transmission line and can supply immediate energy for logic transmission. The characteristic impedance of this power source can be less than 1 Ω. Unfortunately the extent of this low-impedance source is limited and decoupling capacitors are still needed. If the time it takes a wave to reflect from the edges of a low-impedance source is 1 ns and the voltage may sag 1% for every nanosecond until the energy can be resupplied. If many logic lines are making demands on this source of energy, then the parallel ground planes become ineffective in a short period of time.

3.14 EXTENDED GROUND PLANES

The equipotential ground plane has a low impedance from edge to edge when viewed as a square of material. In theory this square can be of any size. If two ground planes are connected together using a wide strip of copper, the ohms per square principle can be applied. The copper strip can be viewed as a series of squares. For example, if the copper is 1 mΩ per square then a 50-ft run of 1-ft wide copper represents an impedance of 50 mΩ. The current in the two ground planes is not uniform because the strip makes an edge contact. This further raises the effective surface impedance. The surface impedance can only be kept low if the connection is as broad as the ground plane itself. Large round copper conductors are so inductive at high frequencies that this type of connection is of no practical value. It provides no additional electrical safety, as this safety is implicit in the power wiring itself.

Ground planes can be extended between floors in a building. The connection should be as wide as an entire wall. Single conductors are no more effective than the building steel. Cables routed between floors should follow the ground plane.

3.15 BACK PLANES

Groups of logic cards are often plugged into a common mother board or backplane. This board serves to interconnect the cards, provide power, and provide bus or I/O connections. This board is often built without a ground plane as there are no logic elements mounted to its surface. This often means that the tight field control maintained on the cards is lost on the backplane.

If the backplane has a ground plane, it must make a broad connection to the ground plane for each card. This presumes that there is logic being transported to or from the cards. The problem is no different than any other ground-plane extension. If the current is forced to concentrate, there are higher fields and there is inductance. These areas radiate and can be susceptible to external radiation or crosstalk.

The power-supply return conductor does more than supply power. It is involved in the way high-frequency energy is transported. If the gound plane is not provided, a grid can be formed on the backplane to approximate this surface. It is important to visualize every logic line and its return path to make certain that the loop area is minimized.

Ribbon connections to the backplane should carry this grounding philosophy onto the ribbon. A ground return path should be made adjacent to each logic path. This means every third ribbon conductor should be a ground. This treatment is not always necessary, but this does represent best practice. The grounds must all be paralleled at both terminations of the ribbon cable. The signals will follow those conductors that store the least amount of field energy.

4

RADIATION AND SUSCEPTIBILITY

4.1 RADIATION AND RADIATION COUPLING

The electromagnetic wave that is far from its radiating source is often referred to as a plane wave. This wave consists of both an electric field and a magnetic field. In a vacuum the ratio of the E field in volts per meter to the H field in amperes per meter is 377 Ω. This number is called the wave impedance of free space and is independent of the nature of the source. The wave impedance inside a thin conductor is the conductivity of the material in ohms per square.

The fields near a radiation source are dependent on the nature of the source. A voltage source such as a flyback transformer or a dipole antenna has a high-impedance field near the source, which means that the E field dominates. The H field dominates near a current loop and this is called a low-impedance or inductive field. The crossover from near to far field occurs at about $\frac{1}{6}$ wavelength. At 10 MHz this distance is about 5 m. The near-field wave impedance at a given distance from a radiator is 377 Ω times the ratio of the far-field/near-field frequency to the frequency of interest. For example at a distance of 1 m from a voltage source, the near-field/far-field interface frequency is 50 MHz. At a frequency of 5 MHz, the ratio of frequencies is 10:1. The wave impedance for a voltage radiator is thus 3770 Ω. If the source is a current loop, the wave impedance is 37.7 Ω. This wave impedance idea is used to evaluate the effectivity of shielding materials. In general, when the wave impedance is low, the shielding problem becomes more difficult.

63

4.2 FIELD COUPLING INTO A LOOP

Field coupling between cables, as discussed earlier, was based on the simplifying assumptions that one cable generated a field sensed by the other and that the cables were near each other. When a far field is involved, the concepts of mutual capacitance and mutual inductance are not convenient. The simplest approach is to calculate the E field coupling as if the H field did not exist. Of course the same answer could be obtained using the H field alone, but this calculation is more difficult.

A far field can couple voltage to a conductive loop when the field propagates along the plane of the loop. The coupling is optimum when the E field is in the plane of the loop, as in Figure 4.1. The coupling occurs because at any moment the E field differs in intensity at the two ends of the loop. In Figure 4.1 the field intensity is 10 V. If the height of the loop is 0.1 m, the voltage difference at one end of the loop is 1 V and at the other end is −0.1 V. The difference in voltage is 1.1 V, which is the induced value at this moment in time. The maximum voltage occurs when the loop height or length is one half-wavelength. Based on this mechanism, the coupling increases proportional to frequency until this half wavelength is reached. At higher frequencies the coupling falls off to a minimum at one wavelength. If one is interested in a worst-case analysis, the half wavelength figure is used and no assumptions are made regarding a cancellation.

The coupling to a conductive loop depends on angle of arrival and field polarization. In many applications the wave has been modified or reflected from a nearby structure. The worst-case analysis assumes that the wave is not attenuated by these nearby structures. If the design can accept this assumption, the problem has definitely been solved.

The electric field strength far from a radiating structure falls off linearly with distance, not as the square, as is sometimes stated. This means that added distance from the radiator will not be very effective in reducing coupling.

When the magnetic field dominates, as near a current source, the voltage induced in a loop can be calculated using Lenz's Law. The H field in amperes per meter must be converted to the B field in teslas by multiplying H by the factor

$$\mu_0 = 4\pi \times 10^{-7} \tag{1}$$

The magnetic flux is calculated using the equation

$$\phi = BA \tag{2}$$

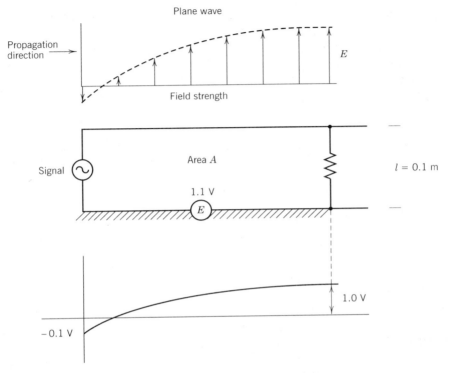

FIGURE 4.1 Field coupling to a loop.

where A is the area of the loop in square meters. The induced voltage is simply the time rate of change of this flux. In the case of a sinusoid this voltage is

$$V = \frac{d\phi}{dt} \text{ rms} \tag{3}$$

assuming H is given in rms units.

The voltage induced into a loop behaves just like an equivalent voltage generator. This voltage source can be used to calculate current that flows in the loop. If this current flow interferes with a normal signal process, the interference must somehow be reduced.

If the offending field is pulse-like in character, the frequency that must be used in calculating the coupling is $1/\pi$ times the reciprocal of the rise time. If the pulse results from an ESD event, the rise time is 1 ns and the frequency of interest is about 300 MHz. In the near field the current level can be assumed to be about 5 A.

A lightning pulse of 100,000 A has a rise time of about 0.5 μsec. This places the frequency of interest at about 640 kHz. The near-field/far-field interface is thus at a distance of 68 m. If this current flows in a steel girder in a building, the near magnetic field is used to calculate the voltage induced into a nearby loop.

The H field 2 m away from a girder carrying 20,000 A is

$$H = I/2\pi R = 1590 \text{ A/m} \tag{4}$$

The B field is about 2 mT. The voltage induced in an area of 0.01 m^2 when the current rises in 0.5 μs is 40 V. This is sufficient to damage electronic circuits.

4.3 SKIN EFFECT

A plane wave reflecting off of a conductor actually penetrates into the conductor. The depth of penetration is a function of conductivity, permeability, and frequency. The wave attenuates exponentially and it is convenient to discuss the concept of skin depth. The depth where the wave is attenuated to $1/e$ or 0.3679 times its original value is called a skin depth. In two skin depths this factor is squared or 0.1353.

A skin depth is given by the formula

$$t = (\sigma f \mu_r \mu_0)^{-1/2} \tag{5}$$

where μ_0 is the permeability of free space or $4\pi \times 10^{-7}$ A/cm, μ_r is the relative permeability, σ is the conductivity, f is frequency in hertz, and t is in centimeters. For copper this simplifies down to

$$t = 6.62/\sqrt{f} \tag{6}$$

At 60 Hz this is 0.855 cm and at 1 MHz this is 0.0066 cm. The skin depth for round conductors is not the same as for a plane surface. The numbers provided by Equation (6) are representative of what might be expected.

4.4 ELECTROMAGNETIC SHIELDING

Conductive surfaces are used to reflect and absorb electromagnetic energy. For a typical metal surface the primary shielding mechanism is reflection. For plane waves or high-impedance near-field waves, a metal surface is an excellent barrier against field penetration. The most difficult fields to shield against are the near-magnetic fields. Magnetic materials are an improvement over copper or aluminum but still offer a limited amount of shielding. As mentioned earlier, it is often simpler to use magnetic materials to redirect the magnetic field rather than try to shield it directly.

Electromagnetic shielding can take the form of conductive paint or plating on plastic surfaces. These surfaces are often very thin and some of the field can penetrate. The reflection factor is most easily viewed as a transmission line mismatch. The attenuation factor is approximately equal to

$$Z_W/4Z_B \tag{7}$$

where Z_W is the wave impedance in free space and Z_B is the wave impedance in the material. For example if Z_W is 377 Ω and Z_B is 0.1 Ω/square, the attenuation factor is 15,080:1 or 84 dB. In the case of a near field where the wave impedance in free space is 37.7 Ω, the attenuation factor is only 1500 or 64 dB.

Plastic materials can be made conductive by adding conductive filaments or powders to the plastic. The percentages of these conductors must be above a critical threshold level or the shielding will be ineffective. It is important that some means of conductive connection around the perimeter be provided or the shielding will not be usable. This can be a buried copper strip or a series of many small wires.

Shielding materials also include screens, conductive glass, and meshes. These materials are used to limit the penetration of radiated energy through holes and apertures in various equipment configurations.

4.5 APERTURES

Apertures are a necessary part of electronics. They are needed to provide cable entrances, viewing, ventilation, access, and control. Energy can radiate in or out of these apertures. If the metal box were completely closed, the energy would not enter or escape except through the conductive walls themselves.

Apertures are difficult to analyze even in their most simple forms. A rectangular hole in an infinite baffle with a plane wave varying in frequency is very complex. The field pattern near the opening has a great deal of fine detail that is difficult to quantize. The field on the far side of the aperture is no longer a plane wave. In the real world apertures are associated with finite metal enclosures not infinite baffles, and, furthermore, there is hardware loaded into the enclosure. An analysis of the actual field that penetrates an aperture and couples to a given internal geometry is thus not very simple.

The only practical approach to this problem is to make a set of worst-case simplifying assumptions:

1. Assume that the field that penetrates is no greater than the maximum field that might penetrate an infinite baffle.
2. Assume the penetrating wave is a plane wave.
3. Assume the wave has the same intensity at all points in the box.

As a first approximation a plane wave penetrates an aperture without attenuation when the aperture's maximum dimension is equal to the half wavelength of an impinging plane wave. If the frequency is lower by a factor of 2, the attenuation is a factor of 2. The attenuation factor is thus the ratio of the half-wavelength frequency to the frequency of interest. As an example, an aperture 1 cm × 10 cm has a half-wavelength frequency of 1500 MHz. This aperture attenuates a 10-MHz field by a factor of 150. The 10-cm dimension controls the attenuation. In some equipment this attenuation is insufficient and the aperture needs to be shielded.

4.6   MULTIPLE APERTURES

When a plane wave is reflected from a conductive surface, the *H* field at the surface is not reversed and this implies surface current. When there is an aperture, this current flow is diverted around the aperture and the wave is not totally reflected. In effect, penetration is allowed when there is a free flow of current around the aperture.

When a group of apertures form a closely packed array, the surface current cannot flow independently around each aperture element. The result is an opening equivalent to one of the aperture elements. A closely packed set of ventilation holes will attenuate a plane wave based on the dimension of one of the holes. A seam closed by mounting screws will attenuate a plane wave based on the dimension between any two screws.

This array phenomenon is quite apparent when conductive screens are considered. A screen allows field penetration based on the size of one opening. The size of the overall screen is relatively unimportant. When a honeycomb is considered the story is different. The array of holes has depth and current can circulate around each individual aperture. This means that each aperture can independently contribute to field penetration. Fortunately, the depth of the honeycomb provides the real attenuation factor.

4.7 CLOSING THE APERTURE

Placing a screen over an aperture does not in itself close the opening to radiation penetration. The screen must be bonded continuously to the edges of the aperture to be effective. If this bond is absent, a new aperture is formed around the perimeter of the old aperture. The dimensions of this opening are identical to the original opening and the aperture is just as transparent as before.

Screens, gaskets, and meshes placed over an opening must make perimeter contact with the bulkhead material. This means that these surfaces cannot be painted, anodized, or subjected to oxidation. Contact areas should be properly plated and sealed. The contact area should be under constant pressure to seal out contaminants and maintain a low-impedance contact.

Cathode ray tubes (CRT) pose a serious shielding problem because of their size. The beam writing speeds and the transitions at the CRT surface make this a difficult shielding problem. If a wire mesh is used, the Moire patterns that result are very disturbing. To be effective, conductive surfaces must remove too much light. Other solutions to the CRT problem involve lensing systems, periscopes, and operator isolation.

Very narrow apertures do provide some additional attenuation over a wide aperture. It is better to be on the safe side and assume that the longer dimension dominates and that the narrowness factor is not too significant.

4.8 WAVEGUIDES

Electromagnetic energy can propagate down a conductive cylinder without a center conductor at specific frequencies. The lowest frequency where this occurs is approximately where the half wave-length equals the largest dimension of the opening. There are many modes of transport and at higher frequencies these modes become al-most continuous. The lowest frequency that the waveguide can sup-port energy is called the cutoff frequency. A waveguide processing a signal at a lower frequency is said to be a waveguide beyond cutoff.

The plane wave that enters a waveguide is attenuated by the ratio of half-wave length to aperture opening. If this signal is in the cutoff re-gion, it is further attenuated by the waveguide. This attenuation is ex-ponential, and is given by the equation

$$A = 30 \ l/d \tag{8}$$

where d is the diameter of the opening, l is the depth of the opening, and A is given in decibels. If l/d is 5, the attenuation is 150 dB or about 30,000,000:1. If this is a honeycomb with 100 holes, the resulting at-tenuation is reduced by 40 dB or still 300,000:1. In effect the wave-guide attenuation is so good that many holes can be tolerated.

Seams and ventilation holes in metalized plastic parts can often be made quite deep and this can be very effective in reducing radiation penetration.

4.9 POWER FILTERS

Filters are used to attenuate signals in certain portions of the spec-trum. Power filters are used to reduce the transport of high-frequency energy on the power conductors. On occasion a filter is used to reduce conducted emission out of a piece of equipment.

There are two modes of interference, namely, differential and com-mon mode. The differential-mode noise is usually caused by load tran-sients and the common-mode interference by external field coupling or ground current flow. Common-mode filters should be placed on all power leads or the common-mode content will not be controlled. If the equipment ground or green wire is terminated outside of the equip-ment, then a filter on this lead is not necessary. Filtering one mode and not the other, or filtering some of the leads, is a rather short-sighted approach to the problem. An equipment ground filter is illegal unless the filter is listed for this application.

The placement of a power-line filter is critical to its proper opera-tion. If the power filter is mounted inside the equipment, the conductors carrying noise can radiate into the equipment. The proper way is to mount the filter on the bulkhead so that radiated energy is kept out of the equipment. The filter functions with respect to its outer case. If the filter can is connected by a length of wire to the bulkhead, the inductance of this conductor will reduce the performance of the filter. A proper filter evaluation requires that the signal source, the monitor, and the filter all be bonded to one ground plane. If the filter is not used this way, it cannot meet its specifications. From this discussion it is obvious that filters located at a building entrance may not be very effective.

Filters that plug into a wall outlet are limited to differential-mode filtering. This is the result of having a long lead for the equipment ground connection. If the equipment is not susceptible to common-mode interference, this filtering may be adequate.

Filters associated with computer power centers are mounted to the frame of the transformer. To be effective these centers must be mounted to the zero-reference gound plane used by the facility.

4.10 HIGH-FREQUENCY FILTERS

The term balun comes from the two words balanced and unbalanced. This term is usually applied to a class of transformer used in rf transmission, where a balanced signal is converted to an unbalanced signal. Balun also applies to a filter formed by threading a group of conductors several times through a toroid. These turns form an inductance that limits common-mode current flow. The balun offers no opposition to the normal-mode signals. This type of filtering can be useful where high-frequency common-mode fields are encountered. Large ferrite blocks are also used. The blocks come in two sections and are clamped around the cable.

Ferrite beads are sometimes placed around individual conductors to form a series element in a filter. The permeability offered by ferrite cores at frequencies above 10 MHz is usually under 100. A single turn has very little inductance and can only function well in low impedance (10-Ω) circuits. Beads often limit cross coupling by spacing leads, not by adding series inductance. If beads are used, care must be taken not to saturate them with dc or large low-frequency ac signals. Ferrite blocks and rods are very brittle and are difficult to mount to survive shipment and normal vibration.

5

ANALOG CIRCUITS

5.1 CRITICALITY IN ANALOG CIRCUITS

The general nature of feedback in an analog circuit was discussed in Section 1.16. An interfering signal is amplified by the net gain of the circuit and reduced by the gain preceding the point of injection. Interference that enters at the input is not reduced by any preceding gain and is obviously more critical.

The noise referred to the input (RTI) of an amplifier is usually the noise of the input stage. In the case of an integrated circuit, this stage is a part of the input device. The noise in a semiconductor is a combination of thermal noise, shot noise, and so-called 1/f noise. The random thermal motion of electrons in a conductor or resistor produces thermal noise. The arrival of discrete charges in a current stream results in shot noise. Shot noise is proportional to the current level in the device.

Semiconductors also exhibit noise that increases at low frequencies and is referred to as 1/f noise. The noise per cycle is roughly inverse to frequency. The design of low-noise integrated circuits is not the subject of this book. Designing circuits to realize low-noise performance is however of primary concern.

A circuit design that is limited by the intrinsic noise of the input stage is ideal. If this noise level is to be protected, the interference level that can be tolerated at every point along the signal path is defined. If the output signal accuracy defines the circuit or system performance, the tolerated interference levels may be higher. As an example, a gain-100 amplifier has a 0.1% error limit and a full-scale output signal level of 10 V. The error limit at the output is 10 mV and this is 100 μV referred to the input. The input stage noise might only be 10 μV. This means that the input can accept interference levels that are 10 times greater than the ideal noise level.

There are three factors involved in circuit susceptibility. These factors are gain, impedance level, and bandwidth. The higher the gain to the output, the more difficult the problem. The impedance of a circuit defines the effect of capacitive coupling. The bandwidth defines the sensitivity to all high-frequency pheonomena. The product of these factors defines a degree of circuit criticality. To use this idea correctly, consider an input stage with a gain of 100. If this stage reduces the circuit impedance by a factor of 10, the circuit difficulty at the second stage is reduced by a factor of 1000. It should be obvious where the greatest design care should be taken and the answer is usually the input circuit. A factor of 1000 gives the designer a lot of leeway.

5.2   OPERATIONAL FEEDBACK

One of the basic techniques in general use is operational feedback. This feedback arrangement is shown in Figure 5.1 and involves two summing resistors and a negative gain amplifier. The gain of this circuit is

$$ G = -\frac{R_2}{R_1}\left(1 + \frac{1}{A} + \frac{R_2}{R_1 A}\right)^{-1} \tag{1} $$

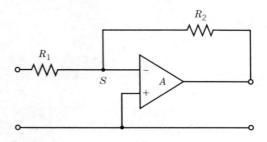

FIGURE 5.1   An operational structure.

When the gain A is large, this reduces to

$$G = -\frac{R_2}{R_1} \tag{2}$$

The departure from this ideal gain is given by the term

$$\varepsilon = \left(\frac{1}{A} + \frac{R_2}{R_1 A}\right) \cdot \frac{R_2}{R_1} \tag{3}$$

When A exceeds the gain by a factor of 10 or more, the error term is very close to

$$\frac{R_2}{R_1 A} \cdot G \tag{4}$$

Stated in words, when the gain A exceeds the closed-loop gain in Equation (2) by a factor of F, the error is simply one part in F. As an example, if the gain is 100 and A is 100,000, this is a factor F of 1000 times the desired gain. The gain error is one part in 1000 or 0.1%.

The point S in Figure 5.1 is called a summing point. The sum of the currents entering this point is zero. This point never moves more than the full-scale output divided by A. In the example just given this is 0.1 mV. If one assumes this point does not move at all, the gain can be determined by inspection. The input current is the input voltage divided by the input resistor. This current cannot flow in the amplifier as the voltage at the amplifier input is zero. The only place for the current to flow is in the feedback resistor. The output voltage is the input current times the feedback resistor or

$$E_O = E_{In} R_2 / R_1 \tag{5}$$

High gains can require high-resistance values of R_2. The circuit in Figure 5.2 provides high gains and yet does not require R_2 to be high in value. The gain can be calculated by noting that the voltage at point P is simply the input voltage times the ratio R_2/R_1. This voltage defines the current in R_3 and R_4. The voltage at the output can now be simply calculated because the current must be supplied by the amplifier. High values of R_2 can limit bandwidth since the parasitic capacitance

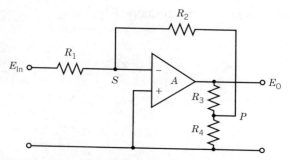

FIGURE 5.2   Operational feedback using a voltage divider.

of the resistor defines the upper-frequency response, which is why this attenuator technique is so valuable.

Any current flowing parasitically into the summing point is interference. In any layout it is wise to keep this node physically small. If the impedances are high, then leakages on the board must be considered. Summing points can be built with ground rings so that leakage currents from power-supply voltages can be guarded out. In critical applications the summing point is not mounted on the board, but up in the air.

5.3 POTENTIOMETRIC FEEDBACK

Feedback structures that compare two voltages at the amplifier input are often called potentiometric. A typical circuit is shown in Figure 5.3. The closed-loop gain can be calculated by finding the output voltage that forces the two input signals to be equal. The gain error is the open-loop gain in excess of the closed-loop gain. In Figure 5.3, if

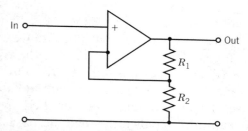

FIGURE 5.3   A potentiometric feedback circuit.

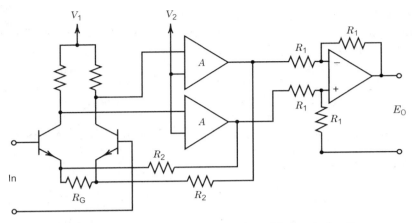

FIGURE 5.4   A differential potentiometric input circuit.

the open-loop gain is 100,000 and the closed-loop gain is 100, the gain error is one part in 1000 or 0.1%.

   The input impedance of a potentiometric system is generally high. If there is an impedance between the two inputs, there is little current flow because the two voltages are nearly equal. One of the most often used low-noise input configurations is potentiometric in nature. The input stage is a matched pair of transistors. The feedback circuit returns signals to the input emitters. The input differential signal is connected to the two input bases. As the bases move, the emitters follow very accurately and, owing to signal processes, the base current is very low. The effective input impedance can be kept above 1000 MΩ. This input circuit is shown in Figure 5.4.

5.4   NOISE IN RESISTORS

Every resistor is a thermal noise generator. The noise is random white noise and is given by the equation

$$V^2 = 4kTRf \tag{6}$$

where k is Boltzman's constant and is equal to 1.374×10^{-9} J/K, K is the temperature in degrees kelvin, and f is the bandwidth. For example, a 10,000 Ω resistor in a 1-MHz bandwidth has a noise figure of 12.8 μV rms at room temperature (300 K). Metal film and wire-wound resistors come very close to meeting this figure. Carbon resistors are much noisier, particularly when they carry current.

The noise generated by a resistor can be replaced by an equivalent series voltage source. When several noise sources are combined, the total noise is calculated by considering the noise from each source separately. The individual voltages are totaled by taking the root mean square. In the feedback structure of Figure 5.1, if $R_1 = R_2 = 10,000 \Omega$, the noise from each resistor is 12.8 μV. The total output noise is thus 12.8 $\times \sqrt{2}$ or 18.1 μV rms.

The voltage divider in Figure 5.2 makes it possible to have $R_1 = R_2$ and still have high gain. In this configuration the noise contributions from R_1 and R_2 are equal. If the voltage divider is designed so that R_1/R_2 is greater than 3, the noise contribution from R_2 will be less than 10% or 1 dB of the total noise.

5.5 TRANSISTOR NOISE

The noise voltage referred to the input in a low-noise junction-type transistor circuit is dependent on collector current. For a collector current of 100 μA, this noise figure is often about 4 nV/$\sqrt{Hz}$. In a 100-kHz bandwidth this is 1.26 μV. For a balanced differential input stage such as in Figure 5.4, there are two transistors contributing to the noise and the figure rises to 1.78 μV rms. This is approximately the noise of a 2000-Ω resistor.

Base current noise is also dependent on collector current levels. For a current level of 1 mA, a typical junction-type low-noise transistor will have base current noise of 1 pA/$\sqrt{Hz}$. This noise has a 1/f character below about 1 Hz. For a source resistance of 10,000 Ω this is an RTI noise contribution of about 3 μV in a 100,000-kHz bandwidth. In an amplifier design it is prudent to keep the input collector current in the input stages low to limit the input base current noise and the input voltage noise.

The noise voltage referred to the input of a typical junction FET or JFET integrated circuit is about 15 nV/$\sqrt{Hz}$. This type of device has input current levels in the pA range. This type of input stage finds application where high-source impedances are encountered. In these applications the source impedance contributes more to the noise than the input stage.

5.6 FEEDBACK AND THE STABILITY PROBLEM

Improper feedback techniques can add considerably to the noise of a system. For this reason it is important to understand how a feedback system is stabilized.

Feedback structures are potential oscillators unless certain conditions are met. The conditions for any circuit to oscillate relate to the delays encountered for a signal to cycle once around the entire circuit. The test for instability must be made at all frequencies. If the feedback loop is broken and a signal is injected at this break point, the signal that returns back to this point can never be greater than and in phase with the injected signal. Figure 5.5 illustrates an unstable circuit because at 100 kHz the signal return is greater than the applied signal.

Feedback to control gain, reduce output impedance, increase bandwidth, and so forth is called negative feedback. For operational feedback the output signal is inverted or 180° from the input signal. This inversion does not represent a delay. In potentiometric feedback the overall voltage gain is positive. In both feedback arrangements the stability criteria is the same. When the loop is broken the returned signal at any frequency can never be in phase and greater than the injected signal. Stated another way, the loop gain $A\beta$ can never reach 180° before $A\beta$ reaches a gain of unity.

Most integrated circuits that provide gain are internally compensated. This compensation controls the phase shift so that the negative input can be tied to the output to form a unity-gain amplifier and the device is unconditionally stable. The phase shift in the open-loop amplifier is usually limited to 90°, although a slightly higher value is acceptable. This means that the gain must fall off proportional to frequency or 20 dB per decade. If the gain were to fall off at 40 dB per decade, the phase shift would be 180° and any feedback would result in an oscillator.

If the circuit does not oscillate it may be marginally stable, resulting in very large peaks in the amplitude response. It is interesting to see what an attenuation rate of 20 dB per decade implies. If an integrated circuit has a gain at dc of one million, this gain must be lost in six

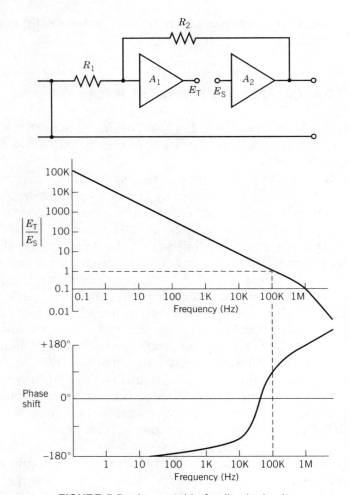

FIGURE 5.5   An unstable feedback circuit.

decades of frequency. If the unity gain bandwidth is 1 MHz, the gain must start falling off proportional to frequency at 1 Hz. If the closed-loop gain is unconditionally stable for gains above 10, then the phase character need only be controlled to 100 kHz.

This loss of loop gain implies that the gain in excess of the loop gain $A\beta$ falls off with frequency. In the example given above the loop gain at 1 kHz is only 1000. This means that the maximum bandwidth for this device at this gain is only 1 kHz. Similarly, the maximum bandwidth at gain 100 is 10 kHz. The relationship between bandwidth and

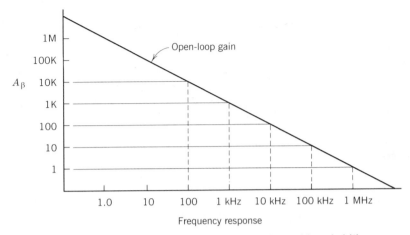

FIGURE 5.6   The relationship of loop gain and bandwidth.

loop gain is shown in Figure 5.6. To achieve a greater bandwidth at
these gains, devices must be selected with greater unity-gain bandwidth.
There are ways to shape the forward gain in an amplifier to limit
the phase shift to say 135° so that there is more loop gain available
at higher frequencies. These designs are more difficult to realize,
but they are effective in providing additional performance in a feedback system.

The need for greater loop gain at high frequencies relates directly to
the accuracies that can be achieved. An amplifier that is running out
of loop gain at 1 kHz cannot perform accurately at this frequency.
When two amplifiers are cascaded to multiply the total gain, the
phase shifts add. An attempt to place feedback around the two gain
blocks will usually create an oscillator because the phase shift is too
great. One very obvious answer is to place feedback around each gain
block separately.

In a design that requires feedback around several stages of gain, the
gain must fall off with frequency on a controlled basis to allow a stable
system. This usually implies losing gain at low frequencies so that the
terminal gain slope is controlled or limited at the upper band edge.
One way to lose gain involves placing a series RC circuit in shunt
across an input collector resistor, as shown in Figure 5.7. The capacitor and collector resistor defines where the roll-off frequency starts
and the series resistor in parallel with the collector resistor defines the

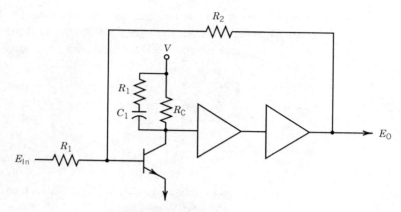

FIGURE 5.7 A gain roll-off circuit.

gain–loss factor that is required. This roll–off has a phase shift of 90° and it must be positioned so that the phase shift including the rest of the system never totals more than 135°.

The essence of the noise problem can now be discussed. When the gain of the first stage is reduced by gain shaping, the noise from the second stage begins to dominate. Remember that the noise injected at any point is reduced by the gain preceding the point of injection. If the input stage has no more gain, then the second stage noise prevails. If the second stage is noisy, the entire device becomes noisy at high frequencies. It is difficult to avoid this problem, for if the gain is rolled off in a subsequent stage, the roll-off capacitor will not allow large ac voltage swings because the stage cannot handle the current requirements.

Improperly designed feedback systems can be marginally stable. This means that there is a peak in the frequency–amplitude response that adds to the noise in this region of the spectrum. Also, every time there is a fast-changing signal at the input the transient character of the amplifier is excited, resulting in ringing. Transients that result in any form of overload produce in-band signals that are in error and can be thought of as another form of noise.

5.7 DIFFERENTIAL AMPLIFIERS

The term differential implies that the input difference signal is to be amplified. If the input signal is referenced to an input ground and the output of the amplifier is referenced to an output ground, the amplifier must somehow still perform. Amplifiers that can function well between two ground references are correctly titled differential amplifiers.

Input signals can be balanced or single-ended with respect to the input ground reference. Input signals can be connected to the input ground reference or they can be floated. Even when floating, there are still parasitic capacitances to consider. Amplifying these various configurations is the domain of the differential amplifier. If the shields are not handled correctly, the signals are easily contaminated.

Consider the output common as the output ground reference conductor. The average input signal viewed from this reference conductor is a common-mode signal. This signal, as shown in Figure 5.8, is simply the ground difference of potential E_{CM}. It has nothing to do with whether the input signal is balanced or single-ended.

If the reference conductor is the input ground, then a part of the transducer excitation can be a common-mode signal. Figure 5.9 shows a strain-gage transducer where one half of the excitation is common mode. This average input signal must also be rejected by the amplifier.

The general rules for electrostatic shielding must apply to the two-ground case. The input shield must connect to the signal conductor where the input signal grounds. The output circuit can be shielded and the shield should be connected to the output common where the output grounds. This latter treatment is not critical but represents best practice.

In general there is a ground difference of potential between any two ground points and this can even exist on the same circuit board. This ground difference of potential is noise in the broadest sense and results

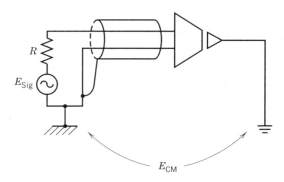

FIGURE 5.8 A typical common-mode signal.

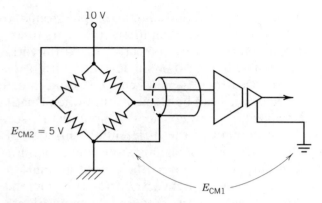

FIGURE 5.9 A second form of common-mode signal.

from field phenomena in the general area. This noise manifests itself in terms of current flow. An amplifier that has one ground-reference conductor forms a ground loop when this reference conductor is connected to two grounds. The current that flows corrupts the signal. A differential amplifier is required in this application to avoid this contamination.

5.8 THE FUNDAMENTAL INSTRUMENTATION PROBLEM

The noise currents allowed in a signal path are a function of accuracy and gain. If the source impedance is 1000 Ω and the signal level is 10 mV accurate to 0.1%, the maximum noise current in the source impedance is 10 nA. If the common-mode level is 10 V, the impedance allowing noise current to flow must be greater than 1000 MΩ. At 60 Hz this is only 2 pF. This number gets ridiculously small if the common-mode frequency is in the kilohertz range. This path for noise-current flow is shown in Figure 5.10, which indicates an undefined coupling between the electronics guarded by the input shield and the electronics driving the output. This coupling can be optical, magnetic, electromagnetic, or conductive. How it is handled is the domain of the instrumentation designer. In most designs in use today there is no guarded amplifier, but merely guarded input leads. The input shield is called a guard shield which must limit leakage (mutual) capacitance out of the guard to less than 2 pF. It is important to note that this

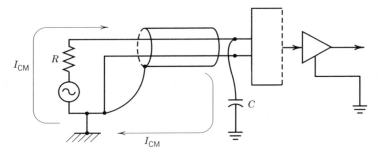

FIGURE 5.10 The leakage impedance in a differential amplifier.

guard shield is not separately grounded at the instrument but at the transducer. A connection to the instrument does not imply grounding.

The electronics located inside the input guard shield must be separately powered. Any power transformer must be multiply shielded to keep power currents from circulating in the input transducer. For this reason most differential amplifiers are built without guarded electronics located inside the input shield. This does not remove the need for a properly treated input guard. Inside the instrument, if there is additional signal conditioning, this circuitry must also be properly guarded.

5.9 THE DIFFERENTIAL ISOLATOR

When the source impedances are low and the gain requirements are minimum, a very simple differential circuit can provide common-mode rejection. This technique, as shown in Figure 5.11, can be used to

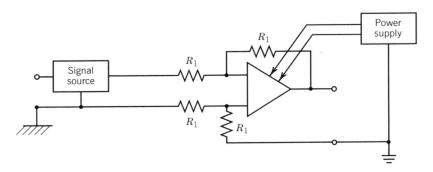

FIGURE 5.11 A differential isolator.

transport signals between two different grounds, where they are not affected by the ground difference of potential. If metal-film feedback resistors are used together with several series potentiometers for obtaining good balance, the ground potential differences can be rejected by a factor of over 10,000 in the band dc to 20 kHz. These isolators must be powered from a power supply referenced to the output ground.

The input signal can be single ended or differential. If single ended, one of the inputs must be connected to the input reference conductor. The choice of which input is grounded defines the sign of the gain.

5.10 COMMON-MODE REJECTION

There are several common-mode signals that influence performance. These signals include ground potential differences, transducer excitation, and power supply variations. The first type is the only one generally specified in instrumentation and here it is usually qualified at one frequency, namely, 60 Hz. There are many other aspects of common-mode rejection that usually go unstated. Because common-mode signals are a form of noise, a review of these areas is important.

1. Common-mode rejection ratios often fall off linearly with frequency. A million-to-one ratio at 60 Hz is often 100,000-to-one at 600 Hz.

2. Rejection ratios vary with gain, depending on how the rejection is obtained. If obtained by an output signal attenuator, the rejection ratio referred to the output is constant. If the rejection is obtained at the input, the rejection ratio is constant with respect to the input. For some rejection schemes the ratio must be provided for each gain.

3. If large enough, common-mode signals can overload an instrument. This overload can result in in-band signals that cannot be filtered out.

4. Common-mode content at high frequency can also overload an instrument or amplifier. This produces offset or in-band signals that interfere with the normal signal.

5.11 DRIFT AND DC OFFSET

Ideally a dc amplifier should always show zero volts out when the input is at zero volts. The offsets and drift are often functions of the source impedance. Some circuits are not designed to accept an open input without problems. We expect an oscilloscope with an open input to display a straight line. As soon as the input leads are exposed to the room, the pattern will be full of 60 Hz and noise. Fortunately there is a position control on the front panel so that the drift that occurs can be nulled out. Many amplifiers require both a connected source impedance and a connection to ground so that the common-mode circuitry is not left floating. Without a definition, the amplifier will drift to some limit or overload. Designs differ and the user must know the restrictions imposed by the design he is using.

The input stage and its design define the drift in a dc amplifier. The usual configuration at the input is a matched pair of transistors. The matching may drift with temperature or time, and this appears as drift in the amplifier. Any input current that flows in the input-active element must flow in the source impedance. If the source is unbalanced, this current will generate an offset signal that appears as a zero error. The lowest noise input stages are designed using junction transistors and the input current is nearly proportional to the collector current. In a good design the current flowing in the input pair of transistors is matched and kept as low as practical based on the bandwidths that are required. A current level of 400 μA is compatible with a 100-kHz bandwidth and drift levels of 1 or 2 μV RTI.*

Assuming that the input stages are properly designed, the way the circuit is laid out is critical. The two input leads must be treated symmetrically along the entire input path. This symmetry is necessary to cancel thermocouple effects at each junction. To avoid temperature differences, the two leads must be kept close together. All dry connections must be through gold contacts and this includes relays. Relay coils that heat contacts in the relay should be avoided. These are precautions that do not cost a great deal and do provide optimum performance. In noncritical applications many compromises are possible.

Air flow plays a role in low-drift amplifiers. Each junction acts as a small thermocouple and ambient air can cause small offsets to occur. Components in the input circuit such as filters or diode clamps should be tested in the circuit to make sure that they are not sensitive to these ambient changes.

Instruments with several gain blocks, where gains in the individual blocks are adjusted, will have both an RTI* and an RTO* specification.

*RTI means referred to input. RTO means referred to output.

Technically a drift specification needs to be supplied for each gain step. The RTO specification obviously dominates at low gains and the RTI specification dominates at high gains.

Drift specifications require a temperature-coefficient specification. To be correct, the input offset, the output offset, and the input current-temperature coefficients need to be specified separately.

5.12 PYROELECTRIC EFFECTS

The amplification of signals from piezoelectric transducers involves so-called charge amplifiers. The charge converter is an operational feedback arrangement using feedback capacitors instead of resistors. The input capacitor is the transducer or crystal and the feedback capacitor in the amplifier defines the gain. A charge converter is an ac device as the reactance of capacitors at very low frequencies becomes unmanageable. To limit the response to say 1 Hz, a high value of resistor is shunted across the feedback capacitor. Input current flowing in this resistor defines the offset of the charge converter.

During warmup or during temperature changes, charges that are bound in the dielectrics are released. The charge converter must provide a path for this current flow through the feedback resistor. If this charge is released too rapidly, the charge converter can overload. The charge converter low-frequency response must be compromised so that the input stage does not block up.

5.13 SIGNAL RECTIFICATION AND SLEW RATE

A diode rectifies by limiting current flow to one direction. Any circuit or device that limits the signal in one direction over another is also a form of rectifier. The differences in performance can be small, but still produce an error. Rectification produces distortion, offset, and apparent changes in gain, all of which interfere with the signal of interest. Rectification is intentionally used to demodulate an amplitude-modulated rf signal. When rectification occurs on an unplanned basis there can be problems. Listed below are a few of the mechanisms of rectification:

1. The simplest form of rectification is signal clipping. This can occur when a signal must exceed the available power-supply voltage.

2. A transistor can only provide current in one direction. If the signal demands a reverse current flow, rectification results.
3. Parasitic capacitances in the signal path require current flow. If this current is not available from the circuit, rectification can result.

It is important to appreciate that a bandwidth specification does not imply that a device will provide full-scale output at maximum frequency. This limitation is usually specified as maximum slew rate. A 10-V peak signal at 100 kHz has a maximum slew rate at the zero crossing of 6.28 V/μsec. If this is the maximum slew rate, the device cannot function correctly at 10 V and 200 kHz. The slew rate is not exceeded however at a 5-V level. Note that the bandwidth for small signals could still be 1 MHz.

Slew-rate limitations can be checked using square waves. In a linear system the settling time is independent of amplitude. The largest signal that can be processed maintaining a fixed settling time defines the slew rate. This mechanism is shown in Figure 5.12, where the maximum slew rate is 5 V/μsec. Slew rates can change based on gain, output loading, and signal direction.

5.14   SIGNAL RECTIFICATION AT HIGH FREQUENCIES

High frequency out-of-band signals can easily be transported into electronics. They are a source of rectification because the circuits do

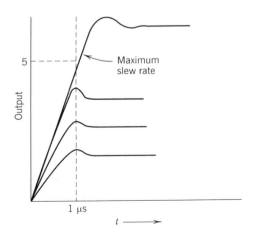

FIGURE 5.12   An example of slew-rate testing.

not have adequate slew rates. Once the signal has been rectified it is injected into the signal stream. If there is feedback, this signal is amplified by the loop gain and reduced by the gain preceding the point of injection. Obviously, the worst case occurs when the rectification occurs at the input. For this reason it is good practice to design analog input circuits with small high-frequency filters.

Plane waves that are reflected or transported on conductors are modified by junctions that have oxidized. This is often referred to as the "rusty-bolt" phenomenon. The surface currents that flow across an oxidized junction are no longer sinusoidal and result in reflected energy that has content at the harmonics of the initial wave. These reradiations can interfere with television or other communications systems.

If an oxidized junction carries power current, the junction is partially open 120 times a second. Any reradiated energy from the involved conductors will be modified 120 times a second. This appears as amplitude modulation on the reflected wave. This is the source of hum appearing on some weak radio signals.

5.15 HUM

The term hum is synonymous with power-frequency coupling. Coupling from the power leads or through the parasitics of the transformer are at the fundamental frequency. Coupling after the rectifiers is apt to be at twice this frequency. Coupling from 400 Hz power is worse than 60 Hz, because both the inductive and capacitive effects are greater. Smaller filter capacitors are required, although the reactive currents are greater.

Most power transformers are built without shields so that the primary and secondary coils are involved in circulating reactive currents in the common conductors of any connected circuitry. This problem was discussed in Section 2.11. If there is a shield, it is best connected to equipment ground to eliminate some of the common-mode coupling from the power line.

The transformer's magnetic field can couple a signal to the equipment. This is not a problem in most digital applications where signal levels are high. The proper approach in layout is to keep low-level signals away from the transformer and to keep the input loop area tightly controlled.

There are many clues available to the observer when hum is a problem. If the hum is associated with the charging of filter capacitors, the current is maximum at the peak of the voltage waveform. If the line voltage is suddenly raised by 5%, the interference amplitude increases until the capacitors are charged to a new value. When the line voltage is dropped, the hum goes away until the capacitor is sufficiently discharged.

The magnetizing current supplied to the primary coil is maximum at zero-voltage crossing. This current flows in the leakage field and is coupled to nearby loops. When the line voltage increases, this coupling will also increase.

Power-line magnetic fields can couple into low-level signal transformers. As indicated before, shielding a low-frequency near-induction field is difficult. The higher the permeability of the core material in the transformer, the less coupling there will be. Nested cans formed from copper and mu-metal can be used around the transformer, but this does raise the cost of the transformer.

Shading coils can be placed around the core of a power transformer. This coil forms a shorted turn for flux crossing the coil. This tends to shape the field so that is more self-contained and directed away from a key area.

Transformers that use primary coils that are wound on split bobbins will have increased leakage flux over a conventional bobbin. These transformers are built to provide safety isolation in applications involving sales in Europe where 220 V is commonly used.

5.16 DISTORTION AND LINEARITY

When a waveform is modified so that the proportions along the wave are changed, there is distortion. The wave need not be a sinusoid to be distorted. To simplify the measurement problem, distortion can be measured by processing a sinusoid through the circuit in question. The distortion occurs once every cycle, and because it is repetitive it is represented by harmonics of the signal being processed.

One form of distortion occurs when an output stage is basically nonlinear and feedback is used to correct for this error. An example might be a stacked pair of output power transistors, operating class B. The current drawn by the stage is near zero when there is no signal to process. The output without feedback would appear to have a dead space around the zero crossings. At low frequencies the available feedback factor is high and the error may be reduced to acceptable levels. At high frequencies the error may be unacceptable.

A distortion analyzer can be used to subtract out the fundamental leaving only the harmonic content, which can be read as an rms value. If the harmonics are singled out one at a time by a filter, the rms content at each harmonic can be measured. There is no simple relationship between the total rms value and the peak value of distortion.

A common way to measure distortion in direct-coupled devices is to measure linearity. When a circuit is linear, the output exactly follows the input. If the circuit responds to dc, the linearity can be measured point by point. For example, if the gain is 2, the input and output will always be a factor of 2 apart. If there is a nonlinearity, this factor might be 1.98 or 2.02 at some point. This particular set of figures represents a nonlinearity of 1%.

Nonlinearity is a measure of peak-to-peak error from the ideal response, and this is not related to distortion. If the peak-to-peak error occurs over a larger part of the cycle, the rms value would be greater but the linearity figure would be the same. The peak-to-peak error need not be symmetrical at the peaks of the signal, and it might even be maximum around the origin. This means that the error must be related to some ideal response. It is necessary for gain and offset to be defined before departures from the ideal can be considered. The following definitions should be used:

1. The output signal for zero input is the reference point.
2. Fit the best straight line through the reference point that minimizes the peak-to-peak error. This defines the gain. Note that this line may not go through the end points.
3. The peak-to-peak departure from this best straight line is the linearity. This definition is shown in Figure 5.13

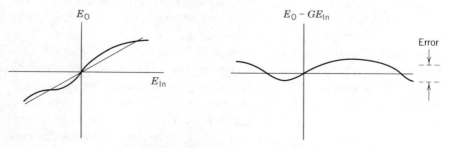

FIGURE 5.13   The definition of linearity.

Linearity can be most easily measured by forming a null network. Here the input signal is subtracted from the output signal. Any residual signal present is the departure from linear. The input or the output signal can be attenuated to form this null. This null observation using sine waves fails when there is phase shift present. Square waves, when properly interpreted, can be used very successfully.

Distortions will generally shift in phase as the testing frequency increases. Since the harmonics are all shifted differently, linearity measurements by a null technique are not possible and the rms measure makes more sense.

A measure of distortion in rf amplifiers is often given in intercept form. The total output distortion power is plotted against input power. Both the signal and distortion are measured in dBm (dB milliwatts). On the same graph the output power is plotted against input power and again both signals are measured in dBm. These plots at small signal levels will be straight lines. The extension of these two lines will intersect on this graph. This point of intersection is a measure of the quality of the rf amplifier. The higher the value of the intercept point, the lower the distortion will be.

5.17 DISTORTION MECHANISMS

Several distortion mechanisms have already been discussed. These include clipping, cross-over phenomena, and slew-rate limiting. Other mechanisms include excess, reactive, and nonlinear loads. When a signal overloads a voltage regulator, the effect can appear as distortion. This overload can also make it appear that hum is added to the signal. When an internal component is overloaded, feedback can obscure this fact, but the loss of open-loop gain will introduce some distortion.

In some systems the feedback loop may open up momentarily. The RTI noise will then appear in the output without feedback until there is recovery. In other systems the circuit may become unstable as the operating conditions change. This instability can appear as "fuzz" or a "spike" on the signal in certain operating regions. Distortions caused by marginal stability are most apt to be observed on high-frequency signals that are near full amplitude.

Amplifiers can sometimes appear unstable because the power supplies are unstable. Bypassing of power supply regulators per the manufacturer's recommendation is mandatory.

5.18   ELECTROSTATIC SHIELDING

The preferred signal cable for low-level analog signals is two-wire twisted and shielded. The shield is often aluminum foil with a drain wire or sometimes braid. The foil has the advantage of providing full optical coverage, whereas the braid has openings. At low frequencies this optical coverage is an indication of the shield effectivity. The foil is not intended for direct connection to a pin at a connector. The drain wire is used because it is far less fragile. This drain-wire cable has limitations if high-frequency signals are to be processed.

The issue of where to connect an electrostatic shield was covered in Section 5.8. To be effective the shield must be connected to the signal, preferably to the zero-signal reference conductor. If one signal conductor is grounded, the shield must tie to this point. A ground here means a connection to earth, a framework, or the low side of another piece of equipment.

In large systems, engineers often want to connect all shields together and tie them to one ground point. This is done without regard to where the signals are grounded. This method has its roots in early electronics, where single-point grounding was adequate. This practice takes place even when the signals are floating. For high-level signals this method may cause few problems, as the shields are probably not needed anyway. For low-level signals where there is limited bandwidth, the shield treatment may also not be critical. This lack of criticality leads to a false sense of security. When there is bandwidth and the shielding is needed, it must be properly handled. Each signal must be shielded separately, without using some simplifying set of grounding rules.

Analog shielding techniques are not effective against radiated emissions (rf). The closure and bonding that are necessary are usually impractical. An rf shield is effective when the field cannot penetrate the shield and this is not related directly to grounding. An rf shield may be multiply grounded along its path. This has the effect of limiting the loop areas so that field coupling is restricted. The field coupled to the inside of this rf shield is proportional to the current flowing in these loops. This is in turn related to skin effect and the quality of the conducting sheath.

Double shielding to limit interference coupling is expensive on a per signal basis and the preferred technique is to provide one rf shield for a group of analog signals. The best outer shield is conduit that is bonded at both ends to available bulkheads.

In most analog designs some sort of input high-frequency filtering is needed. If this is not provided, signal rectification will take place at the input stage. A typical filter might be a series 10-μH inductor, a series 200-Ω resistor, and a shunt 100-pF capacitor. In a differential amplifier, filtering is provided to the input guard shield from both input leads. A bifilar wound inductor can be used to limit the filtering to common-mode interference.

Braid or foil-wrap shields do not provide the basis for good long-line high-frequency transmission. For short runs, a quality shield may not be a necessity. Open twisted-pair wire has been used successfully on 100-MHz busses over distances of 70 ft. Here some sort of signal reshaping is needed. Any of these lines can carry common-mode interfering signals even though the quality of transmission suffers.

The transport of high-frequency common-mode signals into the input of a differential amplifier is unavoidable. The input guard shield cannot be grounded at low frequencies or the common-mode rejection will be lost. The input guard shield must be filtered (grounded) to the output common or the high-frequency common-mode signals will overload the common-mode processes. This filtering should be done outside of the instrument to keep filter current from radiating into the electronics. The best compromise is to filter the input guard to output common right at the input connector. Without this filtering, rectification can result, which causes spurious inband signals. The input filtering for a low-level signal is shown in Figure 5.14.

5.19 LOCAL SHIELDS

The metal structures used to house electronics should be connected to the signal common. It is often a better compromise to connect this

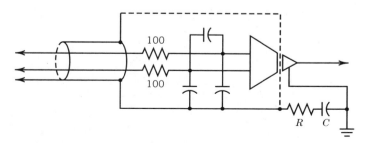

FIGURE 5.14 Filtering and shielding at the input of an amplifier.

metal to a local common rather to signal common via a long input shield run. This is a practical way to avoid coupling high-frequency interference into the circuit.

In some sensitive circuits the local shield must be extremely tight. Power supplies brought into this area with even 1 mV of ripple will capacitively couple noise to the circuit. The charge converter circuit is a case in point. If the feedback capacitor is 200 pF, a 2-pF leakage is a 1% error. If the gain following is 1000, a 0.002 pF leakage capacitance is too large. This requires a tight local shielding.

Exposed metal that might be touched by a user should be grounded in the power sense via the "green wire" to the neutral back in the service entrance. If this is not done, then a fault to this metal will not trip a protecting breaker and it will be a shock hazard. This includes front panels and rack enclosures. This means that signals should not ground to the external cabinet since any other signal grounding forms a path (ground loop) for interference currents to take.

The exposed metal that might be touched by a user could be the discharge path for an electrostatic discharge (ESD). If this metal is grounded through the electronics or via a conductor that goes past the electronics, the radiated field near this path can be very high. This can couple energy into circuits and literally blow them up. External metal should be multiply connected and bonded to all adjacent metal structures so that the ESD path is broad and never constricts the current. If possible, the path should be dressed away from nearby circuitry.

5.20 MAGNETIC SHIELDING

A variety of magnetic materials are available for shielding. One common material is mu-metal. This material has a high permeability even at low flux densities, and is used to shield against the earth's magnetic field and power-related magnetic fields. In one application the metal is formed into a shroud that surrounds a cathode ray tube. This keeps the fields from nearby power transformers from modulating the electron beam.

Mu-metal is formed by annealing the finally shaped product in a hydrogen atmosphere in a strong magnetic field. This aligns the magnetic domains, giving the material its magnetic property. This process limits the physical size of the parts that can be handled. The end product is sensitive to mechanical strain of any kind. Bending, drilling, or dropping the final product will disturb its magnetic properties.

If the metal is heated above its Curie temperature, the magnetic properties will be reduced. Because the material is easily saturated, it should not be used where large magnetic fields will be encountered.

All metals provide some shielding against a changing magnetic field. The mechanism involves eddy currents induced into the metal's surface. For this reason transformers are often shielded against magnetic fields by using nested cans of copper and mu-metal.

High-permeability materials lose their effectiveness as the frequency rises. Skin effect keeps the fields from entering the material and the eddy current losses begin to dominate. This is fine for shielding, but very poor for transformer action. To break up eddy current paths, lamination thicknesses are about 15 mils for 60-Hz applications. At 400 Hz the thickness is about 6 mils.

Powdered-iron alloys can be sintered into various shapes to form transformer cores. The domains of magnetic activity are imbedded in an insulating material and this limits eddy-current losses. This separation of domains also limits the permeability. One common material is known as ferrite and it is used for high-frequency transformers and rf filter elements.

Ferrite blocks are sometimes used to limit common-mode current flow in a group of conductors. An example might be a ribbon cable carrying digital signals. In a few cases ferrite beads can be threaded onto leads to act as a filter element. When this is done the current in the lead should not saturate the material and the impedance of the circuit should be in the range of 10 Ω.

Tapes with a backing of powdered magnetic material are available. These tapes can be wound around a cable to limit magnetic field penetration. The benefit here is mechanical flexibility. These tapes are nonconductive and do not support current flow on the surface; they also tend to divert the field by having the field follow the tape. This material is ineffective in shielding against an electric field.

5.21 CABLES

A large number of cable types are available for the transport of analog signals. Cable types include single-conductor shielded, bundles of two-wire shielded, and groups of 10-wire shielded conductors. Insulations, shield quality, and conductor sizes vary. Cables are often manufactured to order to accommodate a given system.

Cable shield can be braid or foil. With braid, the fineness of the weave can vary significantly. The penetration of external fields into the cable is reduced when the braid is tightly woven. The foil and drain-wire problem has already been discussed. For analog work the foil has many advantages. For digital work the inside drain wire can contribute to interference coupling. A 360° bond to the foil and an external drain wire is preferred, but difficult to execute.

In the general signal-conditioning of low-level signals conductors are required for signal, excitation, calibration, and remote sensing. This can add up to 10 conductors, all located within one electrostatic shield. This shield is usually a guard shield that cannot be shared with another signal cable. When many signals are carried in one bundle the shields must be insulated from each other. An overall shield can be used to limit rf interference. This shield may be bonded to various grounds along the run to limit common-mode coupling. It should be fully bonded around its perimeter to the originating and terminating structures. If practical the outer shield should be a thin-wall metal conduit where the terminating hardware provides a terminating bond. Many cables can share this one path. If braid is used, a backshell termination is recommended. Pigtails that connect the shield to the bulkhead are too inductive to be of any value at rf. It is difficult to let go of the idea that an ohmic connection is sufficient.

5.22   ANALOG PRINTED WIRING LAYOUT

A quality amplifier has the following typical specifications: 100-kHz bandwidth, gains greater than 5000, noise RTI of 2 μV rms, and common-mode rejection ratios of 120 dB at 60 Hz with a 1000-Ω line imbalance. These designs can be made to work on a two-sided board without the use of shielded cable or ground planes. This may come as a surprise to some, but it shows what can be done if care is taken in layout. The basic ideas of analog sensitivity were covered in Section 5.1. The techniques that can be used are straightforward. Guard traces can surround the input leads. The loop area in the input run must be kept small. The components are laid out so that the signal side has the shortest lead length and the least exposure. The signal reference conductor is routed to have the least area with the signal path. Power is routed from output to input to avoid common-impedance coupling. When a compromise is necessary, priority is always give to the input circuit.

Most circuits that are properly laid out will function in the open without special shielding. When hands and fingers get near there may be some rf injection. If the input filters are installed, even this may not cause a problem. This all presupposes that the input circuit is connected to a signal source and is not left floating.

Hand-wired prototype circuits are difficult to build so that they perform the same as a printed wiring board because circuit density is rarely the same. In complex analog layouts, computer-aided design techniques make it practical to avoid most hand wiring. There are added costs and delays but the benefits are significant.

5.23 CONDUCTED EMISSIONS

Analog circuits rarely radiate any significant energy. Circuits that utilize switching-power supplies can cause noise current to flow in the power conductors. These unwanted currents are called conducted emissions. These currents flow in the equipment ground through line filters. In facilities with many electronic devices the proliferation of switching-power supplies has created a severe noise problem. The military often limits the magnitude of this current flow in equipment they specify.

The FCC limits the permitted conducted emissions from equipment where interference could impact other communication devices. Testing procedures for measuring this conductive current are given in the FCC regulations. Typically, a line-impedance stabilization network or LISN is used to make the measurement. The LISN consists of an air-core inductor, a large filter capacitor, and a sensing resistor. The circuit specified is bulky and difficult to build and tests can be made on a modified circuit as long as the measurements are properly interpreted.

5.24 ALIASING ERRORS

When data is sampled on a regular basis, the resulting data can be very misleading. A good example is the motion picture of a spoked wheel appearing to rotate backwards. If the frame rate is high enough, or if the wheel turns more slowly, this problem will not occur.

A person who is allowed to sample the daylight once every day at noon would say that the sun was stationary. A person sampling the light once per day at midnight would not know there was a sun. Both of these observations are wrong and they are both aliasing errors.

If the sampling rate for daylight is once every 25 hours or once every 23 hours then the observer would judge that the day was 24 days long. This is also an aliasing error. All of the phenomena can be seen using a stroboscope to measure rpm. As the wheel spins faster or slower than the strobe rate, the wheel appears to reverse direction. When the wheel stands still, the rpm of the wheel can be determined. If the strobe rate is halved, the observer will see the same stationary pattern and can get confused about the correct answer.

In sampling an electrical signal there cannot be frequency content near the sampling frequency. More precisely, the highest frequency that can be correctly sampled is one half of the sampling frequency. This means that the signal must be low-pass filtered so that there is no frequency content at or above one half of the sampling frequency.

The filters used to remove aliasing errors often have a very steep cutoff characteristic. This allows the user to obtain a maximum bandwidth without error. These filters are relatively expensive, and where there is a choice it is less costly to increase the sampling rate and use a simpler filter. A higher sampling rate requires more memory or a data algorithm that watches for aliasing problems and stores a subset of the data when appropriate. The cost to store data has been steadily decreasing and excess storage is often the preferred route.

The steep cutoff character of the lowpass filters described above have a very rapidly changing phase character near their cutoff frequency. It is very difficult to maintain a phase match between channels of data near the cutoff frequency. The increase in data bandwidth thus comes at the expense of timing accuracy. If groups of signals must be time correlated, then this type of filter may be undesirable. It is possible to correct for this type of error in a computer if the system has been correctly calibrated.

Serious problems have occurred when assumptions are made about the nature of the signal. For example, it would seem safe to assume that a thermocouple has limited signal bandwidth. The problem occurs when noise contaminates the signal. When this contaminated data is sampled the noise can contribute an aliasing error that cannot be separated from valid data. For this reason the engineer must be very careful to examine the nature of the data before he decides to operate without an antialiasing filter.

6

DIGITAL CIRCUITS

6.1 INTRODUCTION

Logic designers often lose sight of the fact that digital circuits are actually high-speed analog circuits in nature. There are rise times, fall
times, source impedances, terminating impedances, transmission lines,
power supplies, and crosstalk. The big difference is in the nature of
the information, which lies in two states of the signal. For TTL a logic
"one" is about 5 V and a logic "zero" is near 0 V.

Wirewrap is still used in some applications where the expense of a
printed wiring board is not warranted. This technique is adequate for
low-speed CMOS and small boards using TTL. Most of the digital circuit boards manufactured today are multilayer printed wiring with one
or two ground planes.

When wirewrap is used, a ground grid should be formed so that
every signal path has a nearby ground return. This can be a grid of
ground conductors or the metal chassis housing the logic. Decoupling
capacitors should also be distributed on a grid across the board. Long
signal leads should be given priority and routed along the ground
plane or next to the conductors forming the ground grid. Direct wiring
is preferred to neat-looking orthogonal wiring. These techniques reduce coupling areas and allow for higher clock rates.

6.2 CLOCKS

A clock is often a square-wave signal that drives the logic through its series of states. Most of the logic in use today involves integrated circuits that are made up of a large number of transistors. These integrated circuits (ICs) are often clock-driven; that is, they perform their function on the leading or falling edge of the clock signal. During the time between clock transitions the incoming logic can be in transition and nothing happens. This means that transients on the logic lines can all decay before the next transition is allowed. Without this technique logic can become unmanageable, or at best very sensitive to cross coupling and to interference.

In systems where the signals are delayed because of transmission time, the clock signals are also delayed. This is apt to be the case where the clock speeds are very high. The designer picks the phase of the clock that is appropriate for that portion of the circuit.

Clock signals must in general have rapid transitions. If the clock is derived from the power line, it will probably be unacceptable. Many ICs will not function on a clock with a slow transition. A Schmidt trigger can be used to shorten the transition time when the speed is needed. Any attempt to filter the clock signal is counterproductive.

Some logic signals arrive at a circuit asynchronously. They must be "clocked" into the accepting logic. The clock system must change states more frequently than the data or errors will result. A variety of algorithms usually imbedded in the interfacing hardware removes any difficulty. Filtering this type of incoming signal can be an acceptable way of removing some forms of interference.

6.3 MECHANICAL LOGIC

Mechanical logic includes toggle switches, keyboards, and relay contacts. There is no such thing as a bounce-free contact. The bounce time is often on the order of milliseconds. Digital logic can take a vote based on a sequence of samples and if the data is not stable it will not respond. Another technique is to process the data, but if there is another transition an interrupt line starts the process over again. This assumes that the data is not momentary, as from a keyboard. It is dangerous to interface logic circuitry from a series of mechanical switches without taking special precautions or providing for the almost certain instability. For example, smart relays with imbedded logic should not be operated from mechanical switches. Of course, if the logic itself is limited to relays, the bounce problem vanishes.

6.4 RADIATION

The rise times associated with logic signals define the nature of the radiation. A repetitive clock will radiate at the fundamental and at harmonics of this fundamental out to a frequency equal to about one third the reciprocal of the rise time. The radiated energy is apt to come from the most active portions of the circuit. In the case of a microprocessor, the memory address and data lines are quite active and are often the culprit.

The transmission lines used by logic are rarely terminated in a characteristic impedance. The reflections and their associated step functions are timed by lead lengths and not by the clock frequency. Since lead lengths vary, the periods associated with these reflections are numerous. This means that the spectrum associated with these reflections is almost continuous. A spectral analysis of the noise radiated from a digital source will appear almost continuous because of this phenomenon.

It is easy to detect radiation from a small digital circuit. A small battery-operated receiver can be held near the circuit. When it is tuned across its band, the spectral nature of the noise will be apparent.

6.5 NOISE BUDGETS

The bit error rates in digital circuitry are often quoted in parts per 10^{10}. A designer of digital circuitry needs to consider every source of error or these numbers are unattainable. Five of the most common sources of error are:

1. The power supply voltages in different parts of the circuit may vary.
2. The logic signals do not swing across the full power-supply voltage.

3. Temperature plays a role in logic threshold sensitivity.
4. Signals cross talk and may not settle adequately.
5. External interference can couple to the logic lines.

A good design considers a budget for each type of error. This way if the errors accumulate, the logic will not malfunction. Allocating the entire error budget to one phenomenon is very risky indeed.

To further complicate the design, many of the error-producing mechanisms can only be approximated and a safety factor needs to be applied for each item in the budget. For example, if 0.2 V is the budget for interference coupling, then a factor of 3 may have to be used to allow a margin for error. This shrinks the permitted interference level to 0.06 V.

When one logic signal must terminate at many points, the logic designer must make sure that the drive has the capability to deliver a logic signal within the budget allowed. The drive capability may have to supply current or sink current depending on the nature of the logic.

6.6 CURRENT LOOPS

A very useful technique is available for transmitting signals between two grounded circuits. The signal is sent as a constant current and sensed as a voltage drop across a resistor. This current source has a very high source impedance and is not sensitive to ground-potential differences. In effect this is a method for rejecting the ground-potential difference, which is one form of common-mode voltage. This circuit is shown in Figure 6.1.

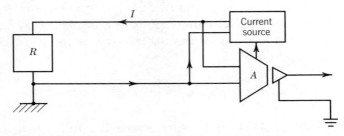

FIGURE 6.1 A constant-current loop for transporting signals.

This method is limited by the characteristics of the cable. At frequencies where the cable can be viewed as a simple capacitance, the high-frequency response is limited. This can be modified by lowering the value of the sensing resistor. The penalty for a smaller resistor is of course a smaller signal.

6.7   BALANCED TRANSMISSION LINES

Twisted pairs of conductors can be effectively used to transmit digital signals with bit rates that exceed 100 MHz. In one technique the signals are balanced with respect to one half the logic voltage. While one line is at a logic one, the other line is at a logic zero. At the receive end a differential logic circuit converts the balanced signal to a single-ended logic signal. This technique rejects common-mode signals that are ground-potential differences. This is a powerful technique in large computer facilities because coax applied to large bus structures would be prohibitive in cost.

When the lines are long, the cable type must be carefully specified. There can be no intermediate taps and the dielectric must not introduce any unusual delay distortion. In some cases the signal being transmitted must be preemphasized so that the cable's characteristics can be accommodated. These long lines must be properly terminated or the reflections will be quite disruptive.

Open lines operating at high clock rates can be very susceptible to pulses from an electrostatic discharge. For this reason cable groups should be transported inside a covered metal raceway.

6.8   PRINTED WIRING LAYOUTS

Multilayer boards are commonplace in digital design. This method allows for ground and power planes, very dense component layouts, high clock rates, reduced radiation, reduced susceptibility to interference, and effective circuit partitioning. When the circuit density is low, circuits must spread out over several boards. The cost of connectors and a second board usually pays for the cost of a multilayer board. The cost and time to hand wire a prototype usually exceeds the cost of a prototype board built in final form. Even if parts of the board need to be redone, the effort to produce the prototype is not wasted. The prototype in this form can actually provide data on radiation and susceptibility which is not available on another layout. Multilayer boards should be laid out using computer-aided tools. The data that results should then be processed by manufacturers that understand this art. Such details as film temperature and humidity affect the registration of holes on the various boards. This level of detail is not for the amateur.

Software is available for automatic routing of connecting traces on a printed wiring board. These tools are valuable provided that the certain initial conditions are defined. Some critical layout parameters must be specified and certain lead-dress restrictions must be stated beforehand. This engineering should never be left to chance. When there is too much reliance on a machine, the results can be disastrous.

The most active portions of a digital circuit should be wired with the shortest leads. This might be a bus connection between a processor chip and nearby memory. Short leads and limited loop area will reduce radiation and increase circuit susceptibility. If there is a choice, slower logic should be located at a distance from the connector.

6.9 THE DIGITAL–ANALOG MIX

One of the key issues in circuit layout is the treatment of grounding at the analog–digital interface. If there is one ground plane over the entire board, it must serve as the grounding point for the analog and digital circuitry. Many argue that this is wrong and that the ground plane must be partitioned into an analog and digital region. It is further argued that if the analog circuitry must be grounded to the digital circuitry, then this connection should be made at one point. This is a little like the single-point grounding rule that is so widely accepted by many engineers. This rule fails, particularly at high frequencies, and other methods are needed to solve the problem.

The problem at the interface of an analog signal and an A–D converter is common-impedance coupling. If the logic current modifies the analog signal, the A–D converter will generate a false logic output. This problem is most serious when the A–D converter has 15- or 16-bit resolution. As an example, if the analog signal is 10 V, then 16-bit resolution is 152 μV. A current of 10 mA and a common impedance of 15 mΩ will produce this level of signal. This is a trace 1 mm wide, 0.03 mm thick, and 3 cm long.

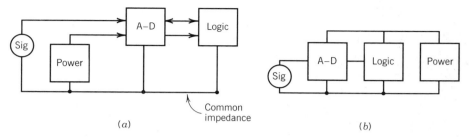

FIGURE 6.2 Common impedance coupling in an A–D converter circuit.

Common impedance coupling can be avoided by using a proper wiring geometry, as shown in Figure 6.2*b*. The logic current flows in a common lead that is not used by the analog input. This approach is simpler than attempting to reduce the coupling by using heavier leads. When high-speed converters are involved the common impedance can be very high, even when a heavy trace is used.

Analog–digital converters are available with differential inputs. This eliminates the common impedance problem described above. The differential input responds to the analog potential difference even though the signal is single-ended.

An analog signal developed on one circuit card is often terminated on a second card. Because the two cards are not on a single common ground plane, there is usually a ground potential difference. This potential difference appears as a gradient along the analog common lead and adds interference to the signal. One way to treat this problem is to isolate the signal using a differential stage, as in Figure 5.11, mounted on the receiving board. The resistors should all be equal and ratio-matched. The common-mode rejection ratio can be kept as high as 80 dB using this approach.

When analog and digital circuitry are located on one PC board, then a ground plane approach without subdivision is preferred. The ohms per square can be below 20 $\mu\Omega$, removing the chance of cross talk from a common impedance. As discussed earlier, a ground plane does not eliminate fields. It only serves to establish an equipotential plane that can limit cross coupling. If the analog areas are separated from the digital areas, then the analog and digital fields will not share the same space. This separation includes components as well as interconnecting leads.

7

COMPUTERS

7.1 INTRODUCTION

The character of computing is constantly changing. The large facility of 10 years ago provided services equal to today's desk-top computer. The large facilities of today have mass memories, printers, magnetic tape handlers, and parallel processors and serve many users in parallel. Small computer installations have been proliferating at a significant rate. A typical installation includes peripherals such as modems, printers, displays, and additional disc memory. Many small systems provide for networking. The power of today's computers has come about because of inexpensive mass memory, the development of large-scale integrated circuits that serve many standard functions, and newer high-speed processors and logic.

The mechanics of radiation and susceptibility are the same regardless of the size of the system. Many of the geometries that are possible on a small printed wiring board are not easily provided when racks of equipment are involved. The physics remains the same, however, and the same principles apply regardless of scale. Big systems must be treated differently only because the physical size requires a different hardware treatment. This point is made because engineers will use a ground plane correctly on a printed wiring board but will fail to make proper use of a computer floor, which is also a ground plane.

109

7.2 SMALL COMPUTER SYSTEMS AND THE FCC

The format for building personal computers is fairly standard. The keyboard and display are housed separately from the main processor or computer. In a lap-top computer the display is an integral part of the main housing. The processor unit consists of a set of cards associated with a so-called mother board. Basically the cards provide an interface for peripherals such as modems, printers, VDTs, and extra memory. These cards have connectors that are accessible from the rear and that accept cables from the peripherals. Power is generally single-phase, 50–60 Hz, and 105–130 V.

Commercial personal computers are required to meet the requirements set down in FCC Regulations 15.1 subpart J. This regulation limits the energy conductively coupled to the power line and the energy radiated out into space. These tests vary depending on the category of the equipment. Personal computers generally fall under category B. The tests are performed under specific conditions so that repeatable results can be obtained in a testing facility. Line filters are used together with a ground plane under the entire test. This regulation is intended to guarantee that the equipment will not interfere with other nearby electronic devices.

The tests performed to meet FCC requirements do not guarantee compliance in every installation. This results because cable dress and equipment location are not controlled. A second factor to consider relates to the use of many different types of peripherals. These tests are simply a base line to guarantee that a product is of sufficient quality. The FCC requirements are not as tight as the VDE (German) or CISPR (European) regulations. The full testing procedure is included with the FCC regulations.

There is no way to separate conductive and radiated emissions. They are interrelated even though the tests are made independently. Radiation can come from the power line just as easily as from the processor card. The conducted noise is measured by inserting an LISN or line-impedance stabilization network into the power line. The radiation is measured using a calibrated antenna located at a prescribed distance from the EUT (equipment under test).

It is expensive to have a testing facility make these tests and provide a certification of compliance. The designer should first test his equipment using a small test probe. This probe can be home built or purchased from a variety of sources. Experience can provide a correlation between the probe results and the FCC test. The FCC itself uses a probe or sniffer prior to a citation because it is easier to use than an LISN and an antenna.

7.3 SMALL COMPUTERS AND MILITARY TESTING

Electronics for military application must often meet a complex set of specifications. Everything from paint quality to component reliability can be covered. Of concern in this book are the issues of conducted or radiated emissions and conducted or radiated susceptibility. These are known broadly by the acronyms CE, RE, CS, and RS. The FCC has no concern for susceptibility, but the military must often operate under conditions of high field strength or poor power conditions. One of the controlling specifications is MIL 461. There are many levels of severity that equipment must accommodate and this is called out in an initial procurement documentation. Often the requirements are made unnecessarily difficult, but this is a different issue completely.

Computers or computer systems that can meet the military's need and operate at high temperatures are both bulky and expensive. Product packaging must provide shielding against external fields. This includes gaskets, conductive housings, CRT screens, special ventilation protection, shielded cables, back shells for connectors, and so forth. These treatments fortunately are effective against both radiation and susceptibility.

7.4 COMPUTER INSTALLATIONS—GENERAL

The term "mainframe" is often applied to computers located in a dedicated facility. These systems are large enough to warrant special power treatment, UPS systems, separate air conditioning, a control console, a communications center, a ground plane, special lightning protection, and so forth. The equipment is located in racks that stand on the computer floor. Cabling is routed under the floor between the various racks.

The ground plane in a facility is a central part of noise control. The ground plane provides an equipotential surface that can be used to reduce common-mode coupling. Cables routed close to this plane will not couple to fields propagating along the ground plane. This control is most effective when the distribution transformers, filters, and racks are bonded to this equipotential surface. Because of its importance, the ground plane needs to be discussed first.

7.5 FACILITY GROUND PLANES

A thin sheet of copper has a resistance below 1 mΩ per square for frequencies below 10 MHz. If a grid of material is used as a ground plane, the bonds at the intersections must have resistances below 400 $\mu\Omega$ or the grid will not be effective. There are many grid forms, including stringers, rebars, copper strips, and various screens. All of these ground plane types can be effective, but in many applications the ground plane is not used correctly and might as well not be present.

The difficulty of bonding two conductors together to obtain bond resistances below 400 $\mu\Omega$ is often not appreciated. In the case of a stringer system, the contact area must be many square centimeters. The contact areas must be specially plated and held together under pressure using spring-type washers. The contact must be treated with a gel to keep the contact area from corroding.

When rebars (reenforcing steel bars) are used, the intersections must be cad welded. This is accomplished using a special chemical burn at the intersection that limits the temperature rise so the strength of the steel is not compromised.

A grid of steel or copper strips 4 in. wide on 2-ft centers makes a good ground plane. The bond at the intersection can be by weld or solder. The entire intersection should connect instead of a point contact such as a spot weld might produce. Sheets of thin steel bonded together by a bead weld can serve as an excellent ground plane. This material is inexpensive and very robust.

7.6 EARTHING THE GROUND PLANE

The ground plane is like any other metallic element in a facility and it must be bonded to the grounding electrode system of the facility. The word grounding has the power definition, which means the conductor is eventually connected to earth. If a hot-power lead touches this metal surface, a breaker should immediately disconnect the power circuit.

The equipment grounding system in any facility provides an immediate and low-impedance return path back to the service entrance. Equipment grounds include any pieces of metal that might come in contact with a power lead if there is a fault condition. Equipment conductors are grounded by the use of "green-wire" or listed conduit. Equipment grounds provide a fault path that must follow the power leads wherever they go. Equipment may be earthed by a separate conductor, but this does not provide for equipment safety. Electrical safety is provided by the electrical equipment itself installed per the National Electrical Code. Note that the impedance through an earth connection is often over 20 Ω and the fault path must be kept in the milliohm range to be effective.

Neutrals and grounded conductors are connected to earth at the service entrance or at the secondary of a separately derived system. These are the only permitted earthing points for power conductors in a facility. The conductor that makes this connection is called a grounding conductor. It is illegal to lift a neutral ground or make a separate neutral grounding connection.

Earthing the ground plane is basic to lightning protection. If a strike current of 100,000 A flows uniformly in a typical ground plane, the potential gradient is kept below 1 V/ft. This means that the ground plane itself is very safe. The current is earth-seeking and therefore the ground plane must be earthed in a low-impedance manner. If this connection is through a single conductor, it will be too inductive to be an effective path. A 3-ft length of #0000 wire has an impedance of 2 Ω at 700 kHz. This impedance implies a 200,000-V drop for a lightning pulse of 100,000 A. It is obvious that the lightning would probably arc to nearby metal rather than follow the single conductor to earth.

The ground plane in a facility must be multiply bonded to building steel to limit potential drops in the region around the ground plane. All metal ductwork, wireways, conduit, and piping in the vicinity of the ground plane should also be bonded to the plane. This is best accomplished by using a ground ring around the ground plane and bonding every entering piece of metal to this ring. The ring in turn is multiply grounded to the building steel. If this bonding is not provided, lightning can be carried by any conductor and arc across to the ground plane. This arcing causes fires and a great deal of electrical component damage. A good rule to follow is that any conductor that comes within 6 ft of the ground plane must be bonded to the ground plane.

7.7 CONNECTIONS TO THE GROUND PLANE

Some engineers argue for a single-point grounding of the ground plane. This is an analog circuit concept that fails particularly at high frequencies. At 60 Hz hundreds of transformers in various pieces of equipment have capacitances from the utility ground to the ground plane. This set of capacitances alone allows 60 Hz and power-noise currents to circulate in the ground plane, and this happens even if the ground plane is earthed once or not at all. In a circuit sense the ground plane has a large capacitance to other nearby conductors including earth. The current flowing in this capacitance is not affected by single-point grounding. It is easy to see that single-point grounding is ineffective and from a lightning standpoint quite dangerous.

The noise frequencies encountered in a large computer facility involve the rise times in the digital logic. Frequencies as high as 300 MHz can be generated. This energy if released can propagate in fields over the surface of the ground plane but not in the ground plane. The presence of a single grounding strap plays no significant role in controlling this field. Filters in this strap will change the field geometry somewhat, but again this is not the way to control the fields that may be present. This type of filter represents a high impedance and will not survive a direct lightning pulse.

Vertical building steel must eventually carry the lightning pulse. This concentration of current implies an inductance and a corresponding voltage drop between floors. The magnetic fields near the steel can be high enough to generate 40 V in a loop (10 × 10 cm) 1 m away from the steel. This obviously will destroy nearby circuitry. This requires that racks of equipment be kept away from any vertical steel.

The edges of a ground plane are not as effective as the center area in providing an equipotential surface. For this reason equipment should be mounted away from the edge. The perimeter of the room can then be used to provide service access to the rear of the equipment.

Ground planes can be extended between floors in a facility provided the connection between floors is as wide as the ground plane. The impedance of a connection can always be calculated by dividing the connector into squares and summing the resistances of individual squares. For example, if the connection is a sheet of metal with an impedance of 500 $\mu\Omega$ per square and the connection is a 100-ft run of 1-ft wide duct, the connecting impedance is 50 mΩ. For a lightning pulse of 100,000 A this is a potential drop of 5000 V. The potential drop for the original ground plane is 50 V. It is easy to see that the duct may not serve to reduce common-mode voltages to levels that would eliminate equipment damage. When a duct must be used as a ground plane, portions of the duct cannot be built in sections and electrically connected using sheet-metal screws. The bond must be as wide as the duct and this implies no paint or anodization at the interface.

Racks of equipment can be made extensions of the ground plane provided that each rack is bonded to the ground plane using a broad strap. Ideally the strap should be as wide as the rack and should make a bond to the wall of the rack, not to a steel member. It may be practical to extend the ground plane into the rack rather than attempt to use the rack structure itself. The ground plane must still be bonded across its width to the rack metal. The cables that exit the rack should be dressed along the ground plane to reduce loop area. If two exits are planned, then two ground planes are required.

Cable dress out of a rack of equipment should follow the ground plane to reduce loop area. This is the correct way to use the equipotential surface that is provided. If the ground plane is a stringer system, the cables should dress along the stringers. This is not done in practice because it is easier to let cables fall to the floor below. This introduces large loop areas between cable runs and the ground plane and it increases the chances of interference coupling.

If the cables that route between rack cabinets are placed on the floor, then this is where the ground plane should be. The stringer system above can serve as an ESD control, to support the cabinets, to form a plenum chamber, and to hide and protect the wiring, but not as a ground plane. The ground-plane extensions that go into the rack should then come from the ground plane on the floor.

7.8   THE RACKS AS A GROUND PLANE

Racks are often bolted together to form a ground plane. Cables between racks are routed along the rack floor or along the rear surface. In theory this method can work if the bonding surfaces are correctly treated. If there is paint or oxidation at the junctions, the bond will not last very long. The result will be a noisy system until all the bolts are retightened.

It is often easier to bolt the racks together but add a ground plane under the racks. The racks are then bonded to the ground plane wherever cables exit. The cables are then routed on the floor on the ground plane. This method does not require special cabinetry and is easy to install.

7.9   THE REBAR PROBLEM

The rebar-type ground plane has several drawbacks. By definition it is buried in the concrete, hiding the nature of its construction and eliminating the chances of adding or changing connections to the grid without breaking up the concrete. Connections once made are single-point rather than broad in nature and this raises the ground-plane impedance. It seems like a good idea to get double usage out of the building steel, but there are too many drawbacks in trying to use the steel as a useful ground plane.

7.10   THE COMPUTER FLOOR

The surface used for foot traffic in a computer facility is usually made from a slightly conductive material. The intent is to dissipate any charge buildup as an individual walks across the floor. If a charge accumulation discharges to the rack of equipment, there could be component damage or a computer malfunction. This discharge is known as ESD or electrostatic discharge. The floor-surface impedance is intended to give a 1- or 2-sec time constant for a body capacitance of 300 pF. The impedance specified in practice can be as high as 10^9 Ω per square.

The stringer system and the covering tiles form a fire-protection barrier. The underside of the tiles must be metal to impede the progress of fire. This metal is not effective as a part of the ground plane unless it is bonded to the stringers around the perimeter of the tile. This is not practical if the tiles are to be removable.

Air conditioning and humidity control are vital to a large computing facility. The ESD problem is significantly higher when the humidity drops below 40%. This is particularly true at high altitudes. Humidity control should be an integral part of the air conditioning rather than a separate unit.

Floor cleaning must avoid dry polishing or rubbing as the potential for ESD generation is too high. Removing tiles often leads to one tile rubbing on another. If the tiles are not grounded when they are moved, the resulting ESD can also be disruptive.

7.11 ISOLATED POWER

The term isolated power implies to some that the power source is somehow floating. The National Electrical Code carefully defines this type of power connection and it is not a floating source. In effect, the equipment grounding conductor or green wire is carried separately from the service entrance, from a separately derived power source, or from a feeder panel to the equipment without connecting to any other conductors. Simply stated it is a dedicated equipment grounding conductor.

The Code very clearly requires that all equipment including receptacles and conduit be grounded to provide fault protection in the standard manner. The only difference is that the equipment plugged into an isolated receptacle is grounded using a separate and usually very long lead.

If the equipment grounding conductor connection is broken or deliberately cut at the equipment, then the equipment can be unsafe. A fault to the equipment housing will not trip a breaker and it poses a shock and fire hazard.

Power-line filters associated with equipment usually mount and reference the equipment housing. A long grounding connection violates the usefulness of this filter as this path is too inductive. The filter capacitors are all in series with a large inductance. The long lead may parallel many other conductors carrying interference and this can couple back into the equipment. If the equipment is sensitive to equipment grounding noise, then the separate conductor idea can be effective, but at high frequencies too many other difficulties can arise and for this reason the practice is questionable. It is far better to design equipment that is not sensitive to equipment grounding. The Code requires that an isolated receptacle be identified by a triangle or delta symbol or the color orange. It is not necessary for the plug to be color coded to match the receptacle.

7.12 RADIATION FROM EXIT OR ENTRY CABLES

Cables that leave an enclosed circuit region can carry energy out onto the cable. This phenomenon is enhanced when the cable passes near an offending circuit. The coupled current flows in all conductors in the same direction and it therefore is classed as a common-mode signal. The preferred construction is to route the cable on the circuit ground plane if it exists. If the enclosure is metal or metalized plastic, the cable should be routed on this conductive surface. This can reduce the coupling somewhat, but the coupling actually takes place near the offending circuit.

Where practical the cable can be filtered at a connector to the enclosure. The connector must be adequately bonded to the enclosure or the filter will not function. Another technique is to surround the cable with a ferrite block to limit the radiated current.

A shield over the cable can be effective if the shield is bonded to the enclosure. If there is no bond or the bond is ineffective, the shield will function as a radiator.

7.13 DISTRIBUTION TRANSFORMERS AND THE COMPUTER FACILITY

It is standard practice to provide separate feeders for a large computer center. This approach provides a certain limited isolation from other user loads. If the computer facility demands are high, it is economical to supply the power on a high-voltage circuit and step the voltage down near the load.

The preferred method of using a distribution transformer for supplying power is to locate the transformer near the computer load. Unfortunately, this transformer is often located near the service entrance in a remote spot and the power is brought in hundreds of feet via conduit. This length of lead is inductive and allows interference to cross couple along the power run.

Distribution transformers that are installed as a part of the facility must be listed for that application and properly installed per the Code. This type of transformer provides what is known as a separately derived power source. The secondary wiring of this transformer must be treated just like a service entrance, except there is no metering. This treatment requires that the secondary be grounded to the nearest point on the grounding electrode system of the facility through a dedicated grounding conductor. A separate earth connection instead of a connection to the grounding electrode system is illegal, noisy, and very dangerous. Floating this secondary is also illegal, noisy, and extremely dangerous.

The preferred location for a computer transformer is on the computer ground plane. A computer ground plane may actually be a good grounding electrode, but because it is a temporary installation, it does not qualify per the Code. This means that a facility distribution transformer must have its secondary grounded to some other proper grounding electrode point such as nearby structural steel. It is very important to remember that a facility cannot have more than one grounding electrode system.

Most listed distribution transformers do not come with internal shields and filters. It is always possible to supply filters separately from the transformers. The only way shields can be added is to interpose another transformer which is not a part of the facility but a part of the equipment. This filter should be mounted on the ground plane or its utility will be limited. Any lead length in series with the filter capacitors adds inductance and the filter will not meet its specifications.

A distribution transformer serves as a filter. The step-down windings lower the source impedance and harmonic currents are filtered by the leakage inductance of the transformer. A short neutral run reduces the common-impedance interference problem that is often encountered.

7.14 POWER CENTERS

Power centers provide distribution transformers designed and listed for mounting on a computer ground plane. These transformers are usually shielded for common-mode interference. These power centers also provide breakers, filters, surge protection, and provisions for grounding. These centers are usually built into a rack that can match other equipment in the installation.

Shielding alone as provided in most power-center transformers is not adequate at frequencies above a few kilohertz. This means that filters may be needed to reduce conductive emissions in the power leads. The filter current flowing in the equipment grounds must be controlled or this too represents a source of interference. This is the reason why the filter return path must be a low impedance as provided by the ground plane.

7.15 ISOLATING EQUIPMENT GROUNDS

The flow of interference current or noise current in equipment grounds is commonplace. A few foolish engineers think that the equipment ground path should be broken to eliminate this current flow. This practice increases the impedance in the power fault path, making the facility unsafe. This practice is not legal per the National Electrical Code. The proper way to control noise in a facility is to recognize how the noise fields propagate and stop them as fields, not by breaking certain paths.

Filters (inductors) in the equipment grounding path are probably illegal since no listed equipment exists. These filters will usually not survive when a lightning pulse enters the facility. The safest approach is to provide a broad, low-impedance path rather that the inductive path provided by a single conductor or conduit. The impedance of a ground plane is less than a milliohm and poses no hazard even for a direct strike.

Wiring on the secondary of a distribution transformer should not fold back into boxes or panels supplying primary power. If this is done, the isolation provided by the transformer and any filtering will be compromised. The equipment ground in the form of conduit must follow all power leads without a break.

In theory filters can be placed inside conduit along the power run. The interference can be stopped from crossing this barrier in the conduit. If the leads that are filtered are paralleled with unfiltered leads, the filtering is violated. Interference that is limited to the inside of a conduit can be radiated when the leads enter an unprotected box or panel. Gasketing these pieces of equipment is not permitted as this violates their listing. All of this indicates very clearly that power filtering should be done at the equipment and the filter capacitors should use the equipotential ground plane for their reference connection.

7.16 MULTIPLE POWER GROUNDING

There are cases where several power sources may enter a facility. The permitted cases are covered by the Code and include different voltages, different frequencies, and multiple sources to satisfy large power demands. Each power source must be treated as a separately derived system and grounded at separate service entrances to the one grounding electrode system of the facility.

The neutrals for each utility power source are grounded, forming a common impedance path. This common path can be the source of interference problems. The problem is minimum when the power entrances are kept close to one another.

Auxiliary or backup power switching poses many problems. The neutrals are often not switched to reduce costs. This means that the grounding arrangements are not optimum when auxiliary power is used. If the equipment fails to perform in this mode, the entire reason for providing backup is lost.

A utility distribution transformer may supply many users. The secondary of this transformer is grounded at the service entrance for each facility. This implies that neutral currents flow in the earth between facilities as not all loads are balanced. This is just one more source of ground potential difference when the earth is used as a reference conductor.

7.17   HARMONIC NEUTRAL CURRENT FLOW

The neutral current for a three-phase load is ideally zero. This assumes that the load demands a sinusoidal current. When the load involves rectifiers, triacs, and switching power supplies, the neutral current can be quite large even though the individual phase loads are balanced. In fact the neutral current can be as high as 1.73 times the line current. If the conductor is not rated for this current, overheating can result and the circuit can be lost due to breakdown.

If three-phase loads are delta connected, a neutral is not required. The neutral must still be grounded at the transformer, but not included as a feeder conductor. This of course eliminates the neutral overcurrent problem and at the same time eliminates the common impedance-coupling problem.

7.18   THE ESD PHENOMENON

The electric field that is present just prior to an ESD is concentrated near the arcing area. Ninety percent of the energy is stored within a

2-ft radius. The discharge energy must come from this space and at the speed of light the energy is dissipated in about 2 nsec. It is thus obvious that the physical system at a distance greater than 2 ft has no effect on the size of the pulse.

A pulse that rises in 2 nsec has frequency content above 300 MHz. Radiation at this frequency is very efficient, so that it is important to consider all apertures in the vicinity of the pulse. Much of the energy will be reflected off of metallic surfaces. Some of the energy will follow along transmission line paths because this is the easiest way to go. To keep the field levels low, there should be several ESD paths and the current flow should never be allowed to concentrate. Apertures should be closed if possible. If not, electronics should not be placed near these openings. Current should never be forced to concentrate on single screws.

8

LARGE INDUSTRIAL
SYSTEMS

8.1 INTRODUCTION

In a typical industrial application the luxuries of a carefully planned computer installation are often not available and the engineer must work out other means of solving the interference problems. The environment can include power transients, motor-speed controllers, fluorescent lights, solenoids, relays, and motor brush noise. To further compound the problem, the transducers can be located over a large area and the cabling can be routed over long distances. Computers can be distributed through a facility and there can be a need to communicate between computing centers. This communication must be reliable even though the environment is very noisy.

8.2 BANDWIDTH

The greater the bandwidth, the greater is the problem of interference control. A thermocouple may have a bandwidth of few cycles, but the amplifier following may have a response of many kilohertz. If the amplifier is overloaded by noise, the temperature measure could be corrupted. This implies that filters should be applied to limit the bandwidth and thus the impact of interference. The filter must not follow the amplification because once the corruption has occurred it cannot be removed.

123

The frequency of interest and rise time go hand in hand. Rise time is defined as the time the event takes to go from 10 to 90% of final value. A transient that rises in 1 μs is characterized by a frequency of 318 kHz. If the rise time is 0.1 μs, the characterizing frequency is 3.18 MHz. An ESD pulse is characterized by a frequency of 300 MHz, a lightning pulse by a frequency of 640 kHz. The very fastest transients are apt to release radiated energy that is transported in free space. The slower transients are more apt to be carried by various conductors acting as transmission lines. It must be remembered that all conductor pairs can transport field energy in both directions. The conductors include shields, conduit, safety wires, control lines, earth, output, and input cables.

Triacs that switch power on twice per cycle in midcycle produce step voltages that can rise 100 V/μsec. The current in a stray parasitic capacitance of 1000 pF is 0.1 A. These currents can propagate through a facility appearing as spikes and pulses and can be quite disruptive. When viewed as a circuit problem, these pulses appear quite unmanageable. They can be contained, but only by using a field view of the propagation.

8.3 INDUCTANCE AND SWITCHING PROCESSES

When current is interrupted in an inductance, the energy stored in the magnetic field cannot be transferred in zero time to some other location. At the moment of interruption, the current representing the magnetic field-energy storage must begin to flow in the parasitic capacitance of the inductor. This capacitance forms a parallel resonant circuit. At the moment of interruption the stored magnetic field begins to transfer energy to the capacitance. For most practical relays and solenoids the natural frequency of this parallel resonant circuit is under 50 kHz. In one quarter cycle all the energy stored in the magnetic field should be stored in the capacitance.

The parasitic capacitance is usually small and the voltage required to store this energy is correspondingly large. What usually happens is that this high voltage breaks the air down in the vicinity of the opening contact and the result is a very fast transient with frequency content in excess of 100 MHz. This fast transient can easily radiate through an entire facility. It has the quality of an ESD pulse.

A typical light switch in a facility will exhibit some arcing when the switch is opened. The magnetic energy is stored in the magnetic field around the current carrying conductors. Fortunately, if the wiring is in conduit, very little energy will leave the confines of the conduit. At the switch plate some energy enters the room, affecting perhaps a nearby radio. If the switch is at the end of some open wiring, the radiated pulse will be much stronger.

The transients associated with a relay or solenoid can be adequately controlled by placing a reverse diode across the coil. When the switch contact is opened, current continues to flow in the diode and the stored energy is dissipated in the coil resistance. This continued flow of current extends the drop-out time of the magnetic device and in some cases this can be undesirable. One example is the use of relays in telephone switching. The operate and drop-out times are critical and a diode cannot be used.

When relay contacts exhibit this arcing they can be placed in a metal box to reduce radiation. Unfortunately, if the leads entering and leaving the box are not filtered to the box, the box will be ineffective. This includes the relay control leads, contact leads, and the power supply leads. From this comment it is easy to see why a telephone office or a communication switch can be a very noisy electrical place. A mix of solid-state electronics and mechanical relays can be difficult to control.

8.4 GROUND POTENTIAL DIFFERENCES

The various equipment grounding points in a facility will all differ in potential. The potential differences include the power frequency and its harmonics as well as various noise signals. Every piece of equipment in a facility has capacitances to the grounding system. This includes motor windings, wiring inside conduit, passive filters, transformer shields, and fluorescent lights to mention a few. These capacitances allow currents to circulate in the grounding system and this represents a source of interference that must be dealt with in any installation. At power frequencies it may be easier to view this phenomenon in a circuit sense, but in truth it is almost entirely an electric field phenomenon.

The multiple grounding of neutrals for separated facilities using one distribution transformer or for several building on one service entrance allows the flow of some neutral current in the earth. The earth is usually high-impedance compared to the conductive path provided in the wiring. The neutral voltage drops can often be greater than 1 V.

A common source of current flow in building steel comes from the installation of distribution transformers. The leakage flux near the core is sufficient to induce current flow in nearby mounting steel. This current can find parallel paths through an entire steel structure. To avoid this problem, all mounting steel in the vicinity of the core should be insulated so that conductive loops are not formed.

The line filters found in many pieces of electronic equipment are mounted on equipment ground. If this equipment is bonded to the grounding electrode system, filter current flow can add to ground potential differences. This problem is most severe when switching regulators are employed in the designs and has become severe enough in some military applications where the current must be limited by a specification. In some installations the ground-current level at 50 kHz has been observed to be greater than 10 A.

With few exceptions it is wise to accept the premise that there will be ground-potential differences in a facility. Instrumentation should be designed to function in the presence of these signals. The techniques for handling this situation are many and this is the domain of the instrumentation designer.

The presence of ground-potential differences in a facility has prompted engineers to interconnect ground points with large conductors. This practice is often futile because the problems are field related, not circuit related. A voltmeter placed between grounds will indicate the fields that are present and the ground tie does not remove the field. Usually there is a nearby steel beam present, and if this is ineffective in reducing the ground-potential differences, the added conductor will probably have little effect.

Ground potential differences can be observed on an oscilloscope. The path of the connecting leads will affect the signal measured at high frequencies. The signal often represents a signature of the electrical activity in the area. The signal difference with and without the fluorescent lights can identify this source of noise. Starting a motor can identify the noise content associated with that motor.

The earth's conductivity varies with terrain, climate, time of year, and moisture content. The grounding rules for safety require a contact with earth of under 25 Ω. This is a low-frequency concept and is usually met by burying a listed piece of metal in the earth at the service entrance. Its primary role is to provide a low-inductance path to earth for lightning so that the surge path does not enter the facility. If the contact exceeds 25 Ω, the rules require a second connection. Long grounding conductors to a remote water pipe are too inductive to be of any value. Water pipes must be a part of the grounding electrode system, but they cannot be the main grounding electrode.

When lightning hits the earth the potential gradient can exceed 1000 V/m at a distance of 100 m from the strike. Buildings separated by 100 m can have ground-potential differences of greater than 20,000 V. It is obvious that surge-protection devices are needed to protect equipment serviced by interconnecting cables.

Conduit associated with any cable entry should enter near the power entrance to keep ground-potential differences under control. Telephone line surge protectors should be placed near the power entry if practical and a bond made to the nearby grounding electrode system. Again, long runs are too inductive to be of any value.

8.5 THE SINGLE-POINT GROUND

The power system makes use of a single-point ground by grounding the neutral once at the service entrance. This method insures that a fault condition in the facility will be detected and will cause a circuit interrupt.

Single-point grounds can be effectively used in analog designs where the circuits are confined to one piece of equipment. The idea is to avoid common-impedance coupling. In a facility where the equipment is mounted at various remote points, the very idea of a single-point ground is impossible.

When multiple signal paths are required, a single-point ground system implies a large loop area for every signal path unless the signal follows the ground connection. This is shown in Figure 8.1 where the loop area is ①, ②, ③, ④. This large loop area will couple to any field in the area and it will be a noisy system.

Single-point grounding ideas are often applied to signal shields alone. This is incorrect in that the proper shield potential is the signal common, not some average ground point. This type of wiring plan is apt to be noisy as the shield couples some alien potential into each signal path.

The terms "good ground" or "best ground" are not based on sound engineering. They are an expression of frustration when the system is

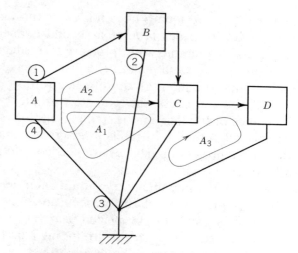

FIGURE 8.1 Single-point ground problems.

noisy and a "better ground" is supposed to "fix" things. Along with these ground types are the descriptions analog, digital, clean, secure, military or NASA ground. None of these expressions have any basis in the schooling provided to electrical engineers. They are an invention that serves an emotional need and have no basis in physical fact.

These special grounds are supposed to somehow "drain" noise currents to a sump in much the same way sewage is disposed of. Unfortunately, an isolated conductor carrying current is a radiator. It also serves to couple to field energy and transport this energy back into the equipment. Circuit theory says that current flowing in a conductor must return back to the source in a loop. The current flowing into ground must somehow come back out of the ground. In other words, the current is not lost in the sump. An isolated conductor is highly inductive and a 10-ft run of #0000 wire looks like 100 Ω at 1 MHz. Even if the ground is buried in salt brine to a depth of 1000 ft, the inductance remains. The ideas presented by people wanting special grounds are specious and have no basis in fact. There are solutions to the noise problems, but they do not have anything to do with these special grounds.

The computer ground plane is an equipotential surface and it works even in an aircraft flying at 30,000 ft. Unfortunately, the earth is a poor ground plane as it looks like a good ground plane perhaps 100 ft under the surface. In the desert or on a lava pile the depth may be thousands of feet. In a building with some reenforced steel and some water pipes the ground plane is too haphazard to be of significant use. This is because bonding to and routing along the conductors is not practical.

Grounding conductors, special ground points, and single ground points cannot be guaranteed to function. An available ground plane is often very impractical and the earth itself is not of much use. This means that both the ground-potential differences and the fields in the area cannot be eliminated. This then implies that the problem of interference control must rest in the way the system is designed.

8.6 THE METHODS OF INTERFERENCE CONTROL

The methods of handling signals vary depending on five factors. These are source impedance, bandwidth, error level, the length of the cable run, and the interference level. Logic, for example, has a high error threshold, a low source impedance, and high bandwidth. A thermocouple has a low error threshold, a low source impedance, and a low bandwidth. A strain-gage bridge has moderate source impedance, low error threshold, and a moderate bandwidth.

There are many overlapping aspects in a design, including economic factors. For example, optical transmission might solve every problem, but the cost might be too high. A high-quality coax cable might be required on a long cable run, but for a short run a simple braided cable might suffice. A universal amplifier might handle all signals, but the cost to handle logic and thermocouples in the same design is too high. A designer must treat each signal separately, looking for commonalty where he can. If there is a saving in using one cable type, then this is a part of the solution.

The designer must select compatible hardware. If the recorder provides no differential isolation, it must be changed or separate isolating drivers provided. If there are no filters provided on a wideband instrument, this signal conditioning may have to be provided elsewhere. If the interference level is known to be high, the equipment must be designed for a high-ambient tolerance. This tolerance must be extended to the cable and to the connector. It is bad engineering to buy good equipment specifications and then misuse the equipment by improper cable or connector selection.

8.7   A FEW MILESTONES

Twisted-pair cable in a straight run can be used to transmit quality colored video signals over a distance of 5 mi. This requires pre- and post-emphasis plus gain. A balanced repeater is needed at every mile point. The receiving and driving amplifiers must be differential in nature to reject power-frequency common-mode signals. A quality video signal requires a 4.5-MHz bandwidth and linear phase out to perhaps 12 MHz. The dielectric used to insulate the wire is critical. Not all insulation materials will perform this well.

Twisted-pair cable can be used to transmit digital signals for 100 ft with a bit rate of 100 MHz. This requires balanced transmission and signal reconstruction at the receiver end.

Two-wire twisted cable with braided shield can be used unterminated to obtain 20-kHz bandwidth over a 3000-ft run. This is differential operation with the shield functioning as a guard shield tied to ground at the transducer. These signal runs are generally unterminated and driven from a near-zero source impedance.

8.8   SOME BASIC TOOLS

1. *Preamplification at the source.* This serves two purposes. The signal level can be raised, thus improving the signal-to-noise ratio and the source impedance can be lowered. The disadvantages are the cost of powering the source electronics and possibly locating equipment in a hazardous area. The excitation can also be supplied from the receiving electronics.

2. *Current loop excitation.* The current for operating the transducer or the transducer preamplifier is supplied from the receiving amplifier. The signal is returned on the same power leads. This method is applicable when the signal of interest has no dc component. The current source can be grounded at the transducer. The receiving amplifier must be differential to reject the common-mode signal. The cable can be single- or two-conductor shielded.

3. *Differential instrumentation amplification.* The input cable can be two- to ten-wire shielded and can be used to handle strain gages, thermocouples, RTDs, potentiometers, and even active devices. The signal can be conditioned in the amplifier. Conditioning includes filtering, balance, offset, calibration, substitution, monitoring, and gain.

Bandwidths to 100 kHz, common-mode levels up to 300 V, and gains to 10,000 are available.

4. *Conduit.* Cables routed in conduit can be protected against external rf fields. This requires proper treatment of the conduit terminations to eliminate field penetration at these points. Many signals can share this one protective sheath.

5. *Isolation transformers.* Located at the signal receiving rack so that the secondary neutral can be regrounded to control the circulation of secondary noise. Provides shields and filtering so that the primary does not couple unwanted common-mode energy to the secondary.

6. *Optical transmission.* Electronics located at the source multiplexes the data and transmits it in digital form over an optical link to the receiver. This method can be economical when bandwidth requirements can be met and the dimensions of the system are great.

7. *Ground planes.* Racks of equipment properly mounted to a local equipotential plane so that local processes are protected against electromagnetic radiation and do not radiate themselves.

8. *Optical isolation.* Digital signals can be optically isolated so that common-mode interference is rejected.

9. *Filters.* Power-line filters to control the flow of interference at high frequencies on the power line. This filtering can control the flow of energy from the load or to the load on all lines including the neutral. To be effective a filter must reference an equipotential surface.

10. *Structures.* The racks and metal works surrounding hardware properly designed to control interfering electromagnetic fields however they originate.

8.9 LIGHTNING PROTECTION

The key word in lightning protection is inductance. The current levels can exceed 100,000 A and the frequency of interest is about 640 kHz. A calculation will show that an isolated lead of #10 wire 1 m long has an inductance of about 3 μH. A surge level of 10,000 A would drop 120,000 V along this length of lead. This same voltage would exist if the conductor were #0000 in size. If this conductor takes a sharp turn, the inductance might double and the voltage drop could exceed 240,000 V. This increase in voltage might break down a secondary path and the surge could enter a facility and do damage. This level of voltage explains why there is so much arcing in a structure hit by lightning. The secondary paths established by the pulse reduce the current level to a point where additional arcing will not take place.

The key to building protection is to provide many low-impedance paths for current flow to earth from the top of the building. The use of air terminals and their spacings is covered in the Code. On the roof the best practice is to provide many parallel paths to the down conductors on the side of the building. There should be no sharp turns and a grounding ring should tie all down conductors and all roof-related conductors together. The down conductors can bond to the grounding electrode system, but this should never add any inductance to the down path. It is not uncommon for side flashes to reach 6 ft. If bonding is not provided, the down conductors should be spaced more than 6 ft away from the building steel.

8.10 SURGE SUPPRESSION

Surge protection is a facility problem as well as an equipment problem. If the only surge suppression is located at the equipment, then the energy levels are apt to destroy the suppression devices as well as the equipment. Even if the surge protection is effective, the magnetic fields near a significant pulse of current can damage nearby equipment. It is preferred that surge protection be provided in stages so that the bulk of the energy is absorbed before it gets to the equipment. This implies that some of the protection should be at the service entrance and at the last feeder panel as well as at the equipment.

Surge suppressors are difficult to design and handle. It is recommended that these devices be purchased as a product rather than built up by the user. They are usually made up of zeners, inductors, and gas-discharge devices. The gas-discharge device can handle the energy, but it is slow. The inductor slows the pulse down so that the zener can absorb the energy on the leading edge. Surge-suppression devices are usually a part of a computer power center when purchased as a unit. Surge suppressors must be properly mounted and bonded or they cannot function properly.

The grounded shields in a distribution transformer provide additional lightning protection. Without a shield, a primary voltage breakdown will follow the secondary leads into the equipment. With a shield, the breakdown takes a direct path to the ground plane.

Cables entering a facility from a remote site should be located inside of a conduit or in a tray that is bonded to the facility-grounding electrode system at the entrance. Outside of the facility a ground wire should be run in air over the conduit or tray. This wire should be periodically earthed along the cable run to provide a down path for lightning current.

When a facility is located at the end of a power run, a good practice is to take the power from the next-to-last pole. A voltage pulse doubles at the last pole as the lightning-induced wave reflects. This extra voltage can then serve to break down a surge suppressor at the last pole carrying current to earth. If the facility takes its power from the last pole, the breakdown current flowing in the earth near the facility can be more disruptive.

8.11   EDDY CURRENT HEATING OF CONDUCTORS

Magnetic fields associated with current flow in power conductors cause eddy currents to flow in nearby metals. At 400 Hz it is generally advisable to use aluminum conduit or ducts to reduce this effect. The magnetic field is reduced when the loop areas involved in power transport are kept to a minimum. A slow twist in groups of conductors will limit the spreading that occurs when conductors are treated singly. This twisting will also reduce cross coupling to other nearby cables.

The magnetic fields around power conductors routed separately through bushings in a sheet of steel can cause heating of the steel. This heating can be severe enough to destroy the insulation and thus cause a fault condition.

8.12   DUCTS AND RACEWAYS

The National Electrical Code requires that all metal structures be bonded together and treated as equipment. If listed for this application, the duct can serve as the equipment-grounding conductor.

Ducts often act as a path for lightning. If the connections to earth are not frequent, the lightning may jump to conductors in the duct for a shorter path to earth. Ducts can serve as a ground plane, but in a limited manner. The usual sheet-metal screw construction does not provide a low enough bonding impedance between segments. One way to avoid this problem is to lay down a continuous strip of copper or steel along the tray bottom that is bonded to the tray along the path. The ohms per square measure of a long narrow run limits the performance of the duct or raceway.

8.13 GROUND-FAULT INTERRUPTION

Power systems in a facility are constructed so that currents flow in designated conductors. The load current is limited to the ungrounded or "hot" conductors and the grounded conductors or neutrals. Load current should not flow in any other conductor, including equipment ground, conduit, or earth. The sum of the currents flowing out and returning in any circuit must equal zero. This summation is used to detect the presence of fault conditions.

A ground-fault interrupter (GFI) is a magnetic sensor that surrounds an entire circuit. It is usually a magnetic toroid surrounding the power conductors. If there is a net current sensed by this sensor, the circuit is interrupted. This happens when some of the load current is returned through the equipment-grounding conductors or through earth. Ground-fault current interrupters or GFICs are used in bathroom and kitchen outlets where shock hazards are the greatest. Here the current flowing in a human being to earth is sufficient to trip the breaker. GFIs can be applied to an entire feeder or facility to disconnect power in the event of a major fault. Here the current trip level is set high enough so that leakages that result from dampness will not cause annoying disruptions of service.

There is always some power current flowing in the equipment grounds. This current is usually reactive current flowing in parasitic capacitances and in filter capacitors located in equipment. When this current gets too high, the GFIs may not function correctly. These detectors are also used to determine fault locations in complex power systems and this logic can be impaired by excessive reactive current flow.

8.14 FLOATING POWER SYSTEMS

In a few applications floating power systems are permitted. For example, the power systems for naval ships are ungrounded to avoid problems of electrolysis and to make them more fault tolerant. The first fault to ground does not disrupt the system and it can be repaired later.

A floating system is apt to be noisier than a grounded system. Every time a load is changed on the system the parasitic capacitances to ground are changed. At the time of switching there are line transients that propagate through the entire system until these capacitances are charged by the line. Equipment transformers are generally not shielded and this noise enters into the circuitry common from the floating primary windings.

Floating power systems are also found in industry, where an orderly shutdown is needed to avoid damage to production facilities. Electric furnaces and assembly lines are examples of this practice. In the furnace situation, if power is disrupted during a melt, the furnace might be ruined. In an assembly line, if one section stops, there might be chaos. Another application of floating power is in hospitals. Here a floating-power source is used to avoid the possibility of a spark that might ignite anesthetizing gases.

Floating the secondary of an isolation transformer located outside of a rack of equipment is in violation of the National Electrical Code. The transformer secondary can be left floating if the transformer is located inside the equipment rack. This is not recommended because the shields present in most isolation transformers, if properly used, will provide better noise protection than a floating arrangement.

8.15 MOTOR CONTROLLERS

The triacs used in motor control provide voltage to the motor over a fraction of the cycle. This fraction is changed based on the regulation requirement. The triacs usually fire twice per cycle so that no net dc rectification takes place. In most applications, control by turning off full cycles is not smooth enough to function well. The voltage when switched on in midcycle is a step function. The voltage may rise 100 V in a microsecond. The spectral content of these transitions is characterized by about 300 kHz, but there is energy content well beyond 1 MHz. These steep rises in voltage couple capacitively through parasitics into the equipment-grounding system. As mentioned earlier, a capacitance of 1000 pF will allow a current of 0.1 A to flow.

The demand for energy from the power line cannot be met on an instantaneous basis. This energy must come from fields storing this energy in the vicinity of the switch. If the source impedance at 300 kHz is 100 Ω, the voltage upon closure is divided between the source impedance and the motor impedance at this same frequency. This

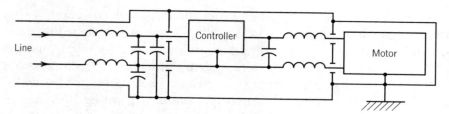

FIGURE 8.2 Filtering a motor controller.

impedance involves the motor inductance and its parasitic capacitance. The power line acts as a transmission line with a step demand for power.

To control this transmission line effect it is desirable to supply the immediate demand for energy from a local capacitor which can be part of a passive filter placed line to line. The voltage that goes forward to the motor will now rise faster and create more problems at the motor. The benefit of a filter is that the step function will not propagate high frequencies backwards on the power line coupling into other equipment.

The forward-going step function should not be slowed down by an inductor, as any practical inductor has a finite natural frequency. If the triac is to be turned on correctly, a high current needs to flow in the triac. Under some conditions a series inductance here can disrupt the turn-on process. A capacitor at the triac switch will guarantee this starting current flow. An inductance can then follow which limits the frequency content of the voltage that is transmitted to the motor. The motor does not need to see a rise time of 1 μsec to operate correctly. In fact, if the rise time is 1 msec, the motor works just fine. This increase in rise time reduces reactive current flow in equipment grounds and greatly reduces the risk of direct radiation from the open coils of the motor. This filtering scheme is shown in Figure 8.2. Only one phase is shown, but obviously all three phases must be treated. This type of filtering should be an integral part of the controller. Retrofitting a motor or controller with filters is obviously difficult. Adding a filter that does not fit the problem will probably not work.

8.16 A NOTE ON SHIELDING

In noisy environments the simplest way to solve the interference problem for cable runs is to filter each signal if possible at the point of entry. Shielded cables are often of little value as proper bonding of the shields at the terminations is difficult and expensive. The inductance added to the shield connection by a pigtail is so high that the interfering field will cause a common-mode signal to appear on all leads. It is also important to remember that fields can enter a controlled box on any lead, including the power, output, control, and even equipment-grounding leads. Filters should be electrically outside of the box, although physically they might appear to be inside. Equipment-grounding conductors should terminate on the outside of the box, not inside.

INDEX